just **THE JOB**

CONSTRUCTION & ARCHITECTURE

COULSDON HIGH SCHOOL
HOMEFIELD ROAD
OLD COULSDON
SURREY CR5 1ES
TEL: 01737 551161

*Also published in the **Just the Job!** series:*

Art & Design
Beauty, Hairdressing & Fashion
Care & Community
Consumer & Home Services
Engineering
Finance & Financial Services
Horticulture, Forestry and Farming
Hospitality, Food, Travel & Tourism
In Uniform
Information & The Written Word
Land & The Environment
Law & Order
Leisure, Sport & Entertainment
Management, Marketing & PR

Manufacturing & Production
Medicine & Health
Motor Vehicles & Transport
Nursing & Therapies
Office & Administrative Work
Scientific Work
Selling, Retailing & Distribution
Teaching
Telecommunications, Film & Video
Working with Animals
Working with the Past
Working with your Hands

just THE JOB
CONSTRUCTION & ARCHITECTURE

Lifetime Careers
WILTSHIRE

Hodder & Stoughton
A MEMBER OF THE HODDER HEADLINE GROUP

> ***Just the Job!*** draws directly on the CLIPS careers information database developed and maintained by Lifetime Careers Wiltshire and used by almost every careers service in the UK. The database is revised annually using a rigorous update schedule and incorporates material collated through desk/telephone research and information provided by all the professional bodies, institutions and training bodies with responsibility for course accreditation and promotion of each career area.

ISBN 0 340 68792 4
First published 1997

Impression number   10 9 8 7 6 5 4 3 2 1
Year   2002 2001 2000 1999 1998 1997

Copyright © 1997 Lifetime Careers Wiltshire Ltd

All rights reserved. No part of this publication may be reproduced or transmitted in any form or by any means, electronic or mechanical, including photocopy, recording or any information storage and retrieval system, without permission in writing from the publisher or under licence from the Copyright Licensing Agency Ltd. Further details of such licences (for reprographic reproduction) may be obtained from the Copyright Licensing Agency Ltd, 90 Tottenham Court Road, London W1P 9HE.

Printed in Great Britain for Hodder & Stoughton Educational, the educational publishing division of Hodder Headline Plc, 338 Euston Road, London NW1 3BH, by Cox & Wyman, Reading, Berkshire.

just THE JOB

CONTENTS

Introduction	9
Bricklayer	13
Carpenter & Joiner	16
Roofing	21
Plasterer	24
Painter & decorator	27
Tiler	30
Plumber	32
Electrical installation & maintenance	35
Heating, ventilation & air-conditioning	38
Fitters and welders. Mechanics. Technicians. Engineers.	
Refrigeration	43
Craft worker. Technician. Professional/chartered engineer.	
Steeplejack	47
Stone cleaner	50
Operating & maintaining cranes & heavy plant	52
Demolition	55
Building technicians, technologists & managers	58
Professional/graduate work in building firms. Technician-level work. Other professional/graduate/technician work in the construction industry.	
Quantity surveyor	65
Architecture	69
Architect. Architectural technicians and technologists.	
Civil & structural engineering	79
Craft workers and operatives. Incorporated engineers and engineering technicians. Chartered engineers. Civil engineering surveyor.	

Road maintenance	88
Wastes management	91
The water industry	95
Naval architecture	98
Shipbuilding & boatbuilding	101

Craft trainee. Engineering technician. Incorporated and chartered engineers.

For further information	105

JUST THE JOB!

The *Just the Job!* series ranges over the entire spectrum of occupations and is intended to generate job ideas and stretch horizons of interest and possibility, allowing you to explore families of jobs for which you might have appropriate ability and aptitude. Each *Just the Job!* book looks in detail at a popular area or type of work, covering:

- ways into work;
- essential qualifications;
- educational and training options;
- working conditions;
- progression routes;
- potential career portfolios.

The information given in *Just the Job!* books is detailed and carefully researched. Obvious bias is excluded to give an even-handed picture of the opportunities available, and course details and entry requirements are positively checked in an annual update cycle by a team of careers information specialists. The text is written in approachable, plain English, with a minimum of technical terms.

In Britain today, there is no longer the expectation of a career for life, but support has increased for life-long learning and the acquisition of skills which will help young and old to make sideways career moves – perhaps several times during a working life – as well as moving into work carrying higher levels of responsibility and reward. *Just the Job!* invites you to select an appropriate direction for your *own* career progression.

Educational and vocational qualifications

A level – Advanced level of the General Certificate of Education

AS level – Advanced Supplementary level of the General Certificate of Education (equivalent to half an A level)

BTEC – Business and Technology Education Council: awards qualifications such as BTEC First, BTEC National Certificate/Diploma, etc

GCSE – General Certificate of Secondary Education

GNVQ/GSVQs – General National Vocational Qualification/General Scottish Vocational Qualification: awarded at Foundation, Intermediate and Advanced levels by BTEC, City & Guilds, Royal Society of Arts and the Scottish Qualifications Authority (SQA)

HND/C – BTEC Higher National Diploma/Certificate

International Baccalaureate – recognised by all UK universities as equivalent to a minimum of two A levels

NVQ/SVQs – National/Scottish Vocational Qualifications

SCE – Scottish Certificate of Education, at **Standard** Grade (equate directly with GCSEs: grades 1–3 in SCEs at Standard Grade are equivalent to GCSE grades A–C) and **Higher** Grade (equate with the academic level attained after one year of a two-year A level course: three to five Higher Grades are broadly equivalent to two to four A levels at grades A–E)

Vocational work-based credits	NVQ/SVQ level 1	NVQ/SVQ level 2	NVQ/SVQ level 3	NVQ/SVQ level 4
Vocational qualifications: *a mix of theory and practice*	Foundation GNVQ/GSVQ; BTEC First	Intermediate GNVQ/GSVQ	Advanced GNVQ/GSVQ; BTEC National Diploma/Certificate	BTEC Higher National Diploma/Certificate
Educational qualifications	GCSE/SCE Standard Grade pass grades	GCSE grades A–C; SCE Standard Grade levels 1–3	Two A levels; four Scottish Highers; Baccalaureate	University degree

INTRODUCTION

There are three sorts of work in the building industry – labouring jobs; semi-skilled and skilled jobs involving work with your hands; and technician and management jobs which may involve working on site but also in an office. There are opportunities ranging from those requiring no formal qualifications to degree level or the equivalent.

Practical jobs

These are the familiar skilled trades, such as bricklaying, plumbing and electrical work, and jobs like roof-slating and scaffolding. In many of the jobs (though not all) you work outside in all weathers. That can be enjoyable in the summer, when there may be a lot of evening and weekend work to get jobs finished. But when the weather's bad, building workers are often laid off. If you have a trade with more scope for indoor work, your earnings are less affected by the weather. Many building workers are self-employed. This means being responsible for your own income tax, insurance and VAT, and covering yourself for holiday and sickness.

You may sometimes have to travel quite a way to work on a big project – to a major industrial site, or to do specialist work on listed buildings. This may mean living away from home for periods, either in this country or abroad. It's possible to make a lot of money that way, and many workers enjoy the experience of travelling around. National Vocational Qualifications at levels 2 and 3 are recognised in many other countries.

The industry has felt the effects of the recession in recent years, and there have been fewer training opportunities. The outlook is brighter now and there is a need for skilled workers in some areas.

QUALIFICATIONS AND TRAINING

It is not essential to have school qualifications, but some GCSEs and a good reference from school can be a help in starting out. The Construction Industry Training Board (CITB) and other organisations offer aptitude tests for people in their last year at school to see what sort of trade would be most suitable. Mathematics, science, and design and technology are generally useful subjects – you need to measure accurately, understand building materials, and read drawings.

Would you like it?

- Even the office-based jobs will involve some work outdoors – whatever the weather.
- For many jobs you need to be fit, healthy and reasonably strong.
- Are you good at maths and able to understand plans and drawings?
- You will need to follow safety rules and regulations.
- Are you able to work as part of a team, but also prepared to work on your own?
- You are likely to have a finished product to look at and think *'I helped to build that!'*

GETTING STARTED

Young people

The main way for young people to get started – for the skilled trades in particular – is through a training placement which will act as the first two years of a three- or four-year apprenticeship. The training covers practical experience and theory, and you

spend time both with a firm and at college. You can gain qualifications up to NVQ level 3. This training is mainly run by the CITB, or by JTL for electrical work (see Further Information section). You may also be able to join a firm as a traditional apprentice in one of the skilled trades, learning on-the-job with block- or day-release at college. A Modern Apprenticeship will provide the experience and training needed to achieve a minimum of NVQ level 3. There are also full-time courses in building trades at colleges of further education, leading to NVQ level

2. After one year of the course, you would still be eligible for a training placement if one was available, or you could continue with a second year.

Adults may be able to train through a government training scheme: contact the local Jobcentre for information. In some areas, women may find special schemes for training in the construction trades.

Technician

The technician is the link person between the architects who design buildings and structures, the site managers, and the skilled craftspeople who carry out the work. Technicians do jobs like estimating, purchasing, surveying and preparing working drawings. This work needs knowledge of the theory and administration of construction. Technician training courses have much theoretical content, and make more use of mathematics and science than craft-level courses do.

Management

Building management is high-level work, responsible for turning an architect's plan into a finished building. A degree or Higher National Diploma in a suitable subject (e.g. building, civil engineering) is usually expected, together with a great deal of practical experience.

Suitable entry qualifications for an HND or degree in building studies (or a related subject area) are A levels in mathematics and physics or a relevant Advanced GNVQ or BTEC National Diploma. A minimum of two A levels (or equivalent) is needed for entry to degree courses, and one A level pass for HND.

For information about other opportunities in job areas related to construction, see also *Engineering* and *Land & the Environment* in the *Just the Job!* series.

BRICKLAYER

Bricklayers do one of the best-known jobs in the construction industry. Bricks are still the main material used in buildings of all kinds, and so there is always a need for good bricklayers. Bricklaying is a skilled job which requires several years' training and experience to become fully qualified.

Safe buildings rely on the basic structure having been put up properly. The way that bricks are laid affects the strength of a building, and so does the type of mortar used to bind the bricks together.

What the work involves

Bricklayers need:

- to work very accurately;
- to work fast;
- a lot of technical knowledge;
- good practical skills;
- to be able to work from plans and drawings.

If they are working on a new building, bricklayers are involved right from laying the foundations to putting the chimney on top. They put up the outside and inner walls – which may be made of bricks, building blocks or other similar materials. Besides putting up new buildings, they can also do maintenance and repair work, and smaller jobs like building house extensions, garden walls or erecting garages. Bricklayers need to know about drainage and scaffolding in order to do their work.

Conditions

A lot (though not all) of a bricklayer's work is done out of doors and, as with most building work, there is plenty of mud and dust around. In the summer, there's often the chance of working long hours and a lot of overtime. In the winter, bad weather can put a stop to work altogether. Building workers are often required to be self-employed, so this can have an effect on earnings. Experienced bricklayers can work abroad, where the pay is often very good.

What it takes

Both males and females can train for bricklaying. You need to be reasonably fit and strong, with good practical skills, happy working at heights and able to put up with dusty conditions. It's very important to be safety-conscious – it's easy to have an accident on a building site – and it helps to like working in a team.

There are no special academic requirements, though you should be of at least average general ability. Maths, science and subjects like design and technology are the most useful ones.

TRAINING

Nowadays, most young entrants start in the construction industry through employer-based training, which acts as the first two years of an apprenticeship. This is largely run by the CITB (Construction Industry Training Board). This means you get on-site experience and go to college or a training centre to get National Vocational Qualifications in bricklaying. There are different levels of NVQ. Employer-based training should help you complete at least level 2 and probably level 3.

Modern Apprenticeships may be available – contact your local CITB office for further information.

CARPENTER & JOINER

> Carpenters and joiners work with wood. Joiners use machinery in a workshop to make items like window frames, staircases and doors to be used on a building site. Site carpenters fit them into the buildings under construction, and do other timber work on site.

What carpenters and joiners do

- Make wooden moulds for pouring concrete;
- cut and fix joists and wooden partitions;
- lay floorboards and flooring panels;
- hang doors, fix skirting boards and window frames;
- erect roof timbers;
- make and install staircases;
- fit kitchen units, built-in cupboards and wardrobes;
- may do decorative panelling.

You have to choose the right timber for a job, cut it to size with a hand or power saw, and fix it correctly into position with a variety of tools. Carpenters and joiners are expected to wear safety equipment on the job, such as goggles, a hard-hat and ear-defenders. You must learn to work quickly from exact measurements on drawings.

To work in carpentry and joinery, you have to be quite strong and fit. You must lift and carry heavy planks and blocks of wood. You work outdoors on items such as roofing and external timber-work, and indoors on flooring, doorframes, skirting, etc. You will need to work at heights.

Where they work

Most carpenters and joiners work on building sites. Some do maintenance and improvements on houses and other buildings. They either work as employees of building firms, or on a self-employed basis. Carpenters can also work for large firms and organisations which have their own construction or maintenance workshops, such as the Armed Forces. There are also opportunities in joinery workshops.

Routine ... or variety?

The work you do depends on the sort of firm you work for and the particular job you are doing at the time. Some jobs mean doing the same thing over and over again. In a big joinery workshop, you may construct hundreds of identical window frames. But if you work for a small general building firm, you may find yourself making purpose-built cupboards and staircases for older housing. You might also turn your hand to all sorts of other jobs, when there isn't any carpentry needed for the moment – putting up plasterboard, for example.

Things to think about

- In the building industry, work is not always in one place. You will need to travel from job to job, so you are likely to need a driving licence.
- This type of work is physically demanding, so it is important that you are fit and healthy.
- Work is not always regular. Like the rest of the building industry, carpenters may suffer when the weather is bad. This might be difficult for people who are self-employed, which, in the building industry, is often the case. Employers prefer to take on workers as they need them, rather than keep them on a permanent payroll. This is where it can be an advantage to have a waged job as a **maintenance carpenter**.
- But there are good points. When the building industry is booming, and the weather is good, there are never enough skilled carpenters. Then, you can pick and choose jobs and make a lot of money, if you are a self-employed craftsman. Again, on the plus side, there is often a good chance of getting overtime if you want it.
- *Prospects:* there are opportunities for setting up your own small business. Skilled workers with experience can become supervisors, estimators and instructors, or go abroad to work on contracts.

Sam – carpenter and joiner

❛ I always enjoyed design and technology at school. We didn't get much experience of using wood, but when we did I just felt it was a material I wanted to work with. I love the smell and feel of it. I didn't want to stay on at school or college – it just wasn't my scene – so I managed to get on a Construction Industry Training Board sponsored placement with a local builder and went to college part-time.

I was surprised at how much maths you need to know to be a carpenter and joiner, but it becomes easier when you're working out something for a practical reason – you can see the logic behind it. I worked my way through National Vocational Qualifications to level 3. The firm I work for does a lot of renovation work, so a lot of the woodwork has to be a one-off job, which is very satisfying.

Your work has to fit in with what everyone else is doing; you can't put a window in before the wall is built up to that point! The brickies, plasterers and roofers all get on well together – we have to! They were a bit surprised to see a female but they know I can do the job, which is what really matters. ❜

GETTING STARTED

Young people
If you are interested in learning woodworking skills, apply direct to the Construction Industry Training Board (CITB). Interviews are held at a local office, when you might be asked to take a simple aptitude test. No fixed qualifications are needed, but it is good to have GCSEs at grade D or above in

English, mathematics, science, computer studies or craft and technology, if you want to follow the full craft training period.

The majority of young people start on a programme of employer-based training (with the CITB). This combines college study with on-the-job experience, working towards National Vocational Qualifications at levels 2 and 3. The training forms the first part of a traditional apprenticeship. Apprenticeships are still offered by some firms; they run for three years with four days a week on site and one day on college release. Modern Apprenticeships may be available – contact your local CITB office for further information.

You may also find traineeships in joinery/woodworking offered by firms which make doors, window frames, etc. These do not provide such a skilled training, and are basically wood machining.

Adults
Adults may be able to train to fully skilled level through a government training scheme. This could be done either through a CITB scheme or through a placement with a suitable employer, plus day release to a college or training centre. Your nearest Jobcentre can provide details.

FINDING WORK

Careers service staff will tell young people about any vacancies for apprentices or trainees, and about suitable opportunities for training in employment. You should write to the CITB, and look out for adverts in the local papers. Try small firms on the off-chance that they might have an opening. People over 18 should also try the Jobcentre for vacancies.

ROOFING

> Roofing is a specialist job in the construction industry. There are many different types of roofing, and firms often specialise in one particular type. After training, it is possible to work for a roofing firm or to be self-employed.

Roof slating and tiling
This is the oldest and most traditional of roofing methods. When the roof shape has been completed in timber, it is overlaid with bituminous felt. Timber battens are fixed on top, set out for the tiles or slates to be fixed upon them. The edges of the roof are bedded and pointed in cement or fixed with edging shapes. The slater and tiler will need to calculate how many tiles or slates will be needed for the job. Slates will have to be cut to shape to fit around chimneys and ridges, and the gullies made waterproof with lead flashing. The job may include stripping a roof of old slates or tiles before replacing them.

Roof sheeting and cladding
Large industrial buildings such as factories or warehouses often have roofs weatherproofed by a covering of flat or corrugated sheeting in various materials such as aluminium, steel, plastic, fibreglass or cement fibre. The sheeter and cladder needs to know the properties of different materials, and how to cut, shape and fix them.

Built-up felt roofing
Flat roofs can be covered by rolls of felt, glass fibre or polyester material, in layers bonded together by hot molten bitumen.

This obviously has to be done with the greatest of care, and the roofer needs to understand things like condensation, ventilation and weatherproofing awkward corners.

Mastic asphalting

This process is used for roads, floors and pavements as well as for roofing. The asphalter spreads the hot material over a prepared surface to the correct thickness, with chippings or paint as the final surface.

What it takes

Roofers need:

- to have a good head for heights;
- to know about building regulations;
- to be able to follow working drawings and instructions;
- to be fit and strong – roofing materials used are very heavy;
- to be happy to work in all sorts of weather;
- to be able to work as part of a team.

TRAINING AND QUALIFICATIONS

None of these jobs require particular qualifications, though you do need to be able to follow written instructions and make calculations. The usual way for young people to get started is through employer-based training. This means learning both on-the-job and at a college or training centre to work towards National Vocational Qualifications at levels 2 and 3. NVQs for the roofing industry include level 3 Estimating, Surveying, Buying and Planning, and level 4 Manager. Training is provided by the Construction Industry Training Board. Training usually lasts two to three years (four for mastic asphalting). Modern Apprenticeships may be available – contact your local CITB office for further information. The organisations listed in the Further Information section can help you to find suitable training.

Roofing technician

Roofing contractors employ trainee roofing technicians, which is more of a supervisory and office-based job (though site work is also involved so a head for heights is still needed).

Technicians specialise in estimating, measuring, costing and drafting as well as supervising and checking the work of craft-workers and operatives. Some academic qualifications will be required, but these vary according to the employer. Training is likely to be on-the-job, with a distance-learning package provided by the National Federation of Roofing Contractors.

PLASTERER

> Plastering is a 'finishing' trade, carried out when the main structure of a building has been completed. It takes several years of training and experience to become qualified, and it is then possible either to work for a firm or to be self-employed.

What the work involves

The plasterer's job is to get the walls, ceilings and floors of a building ready for the painters, decorators and floor-layers. Plasterers mix up plaster and spread it to form a smooth plaster surface on new walls, ceilings and floors. This work is known as *solid plastering*.

In old buildings, plasterers repair and replaster damaged surfaces. They also put up plasterboard partitions and ceilings, and prepare screeds on to which floor coverings can be laid. A knowledge of things like insulation materials is needed.

Plastering is one of the building trades which is mainly done indoors, though half-completed buildings still leave you partly exposed to the weather. Sometimes plasterers work on the exterior walls of buildings, applying a surface finish such as stippled concrete.

There is also specialist work called *fibrous plastering*. Plasterers make decorative plaster mouldings for fixing to ceilings, walls or columns, either in new buildings or as part of a restoration job. The mouldings are produced in a workshop and fixed on-site.

What it takes

Plastering may look easy, but that is probably because you've watched someone who knows what they are doing! Like all building crafts, there are skills and techniques which you only gain by practice. You have to learn how to mix plaster correctly – that is absolutely crucial to doing a good job. And you have to learn how to apply it to surfaces smoothly and evenly.

Plasterers rely heavily on good hand-to-eye coordination, as they have to work at a fairly rapid rate.

This is a job which needs a lot of fitness and stamina. Plasterers use the muscles in their arms and wrists a great deal, and do most of their work standing up. You hold the handhawk on which you carry the plaster in one hand, and use the other to apply the plaster with a trowel or float. Besides large expanses of walls, you have to work in small confined spaces where it can be very awkward. You need a good head for heights and to be a patient, careful worker.

QUALIFICATIONS

You don't need any particular academic qualifications to become a plasterer. Maths is useful for estimating and measuring quantities of materials and areas to cover. Both young men and women can get training. As with all building trades, it helps to be able to drive, so that you can get from job to job easily.

GETTING STARTED

The usual way for young people to get started in plastering is through employer-based training. This means learning both on-the-job and at a college or training centre to work towards NVQs available at levels 2 and 3. Your training might be under the control of the Construction Industry Training Board or another approved training organisation. Modern Apprenticeships may be available – contact your local CITB office for further information.

just THE JOB

PAINTER & DECORATOR

> Painting and decorating is one of the skilled trades in the building industry. It takes several years of training and experience to become qualified and it is then possible to work for a firm or to be self-employed.

Painters and decorators do one of the finishing jobs in the building industry. When a new building has been put up, one of the very last jobs is done by the decorators, who paint it inside and out and perhaps fix wall coverings, ready for the occupant to move in. Painters and decorators also do a great deal of work on older buildings, redecorating and repainting interiors and exteriors. They work on all types of buildings, from small houses to mansions, factories, schools and office blocks.

Painters and decorators can work for building firms, for painting and decorating firms, and with large organisations which have their own maintenance departments. It is also possible to set up your own small business – not so much capital is needed as for some ventures, and you can start on a fairly small scale.

What the work involves
Although many people do their own home decorating without any training at all, doing it professionally is a very skilled job. Work has to be done to a consistently high standard, and you will do jobs well beyond the scope of the home decorator. It's often when the handyperson can't cope, that a decorator is called in!

When painters and decorators are trained, they learn how to work with a wide range of materials – lots of different sorts of paint, and wall coverings, ranging from papers to hessian and other fabrics.

They also have to know how to treat and prepare different surfaces ready for painting or decorating. Cracks and holes have to be filled, and old layers of paint or wallpaper must be removed, before the new ones can be applied.

Painters and decorators have to learn how to use the tools of their trade – all the types of brushes and rollers, and perhaps mechanical hand tools and spray-painting equipment. As with any craft, it takes time to become really skilled at painting and decorating techniques. In all aspects of painting and decorating you need to be fast, clean and accurate.

What it takes

Painters and decorators need:

- to be tidy and methodical workers, with good practical skills;
- to work in a team;
- to get on well with customers;
- to be able to work at heights, for instance when decorating stairwells and the exteriors of buildings;
- to be safety-conscious when working on scaffolding and using chemicals;
- to have good colour vision, and to be able to match colours.

There are no special exam requirements, but it is useful to be reasonable at arithmetic, so that you can estimate materials and costs, and measure accurately.

GETTING STARTED

The usual way to get started in painting and decorating is

through employer-based training. This involves learning both on-the-job and at a college or training centre to work towards National Vocational Qualifications, currently available at levels 2 and 3. Your training might be under the control of the CITB (Construction Industry Training Board) or another approved training organisation. Modern Apprenticeships may be available – contact your local CITB office.

TILER

> A tiler's skills provide finished surfaces that are clean and practical, and often decorative, in all sorts of buildings. No academic qualifications are required to start.

A **wall and floor tiler** works on many different size jobs. A customer may want a tiled splashback for a kitchen sink, or a bathroom tiled from top to bottom. A supermarket, swimming pool, shopping precinct or office block may require a huge area to be tiled. A tiler working for a large contractor may have to travel a long way from home to work on a big project.

There are different types of tiling, too. A job may just involve laying tiles in straight lines across a rectangular area, or making a complicated pattern. Tiles have to be cut to fit round pipes, window ledges and anything else that gets in the way.

What the work involves

The tiler has to mix sand and cement mortar to prepare an even wall and floor surface, using a straight edge and plumb-line. Careful planning is needed to see that the tiles are straight and look right. This usually involves setting the tiles out first, starting at the centre and working outwards. The tiles are fixed on cement or some other adhesive, which is allowed to dry, and then grouting paste is applied to fill in and seal the gaps.

Tiled surfaces may be indoors or outdoors, so you must be willing to work in all conditions and able to work at height. You need to be fairly fit because of the bending and lifting involved.

You should be able to work on your own or in a team with others.

TRAINING AND QUALIFICATIONS

Although no particular educational qualifications are needed, you will need to be able to follow written instructions and make calculations.

The usual route for school-leavers is through employer-based training. This is usually a combination of on-the-job and off-the-job training. Off-the-job training is carried out at a CITB (Construction Industry Training Board) training centre, working towards NVQ level 2. Full information is available from the CITB. Modern Apprenticeships may be available – contact your local CITB office for further information. Your local careers service will be able to help you find training. Adults should enquire at the local Jobcentre.

PLUMBER

Plumbers do all the jobs concerned with the water and drainage systems of buildings. They work both inside and on the outside of buildings of all kinds – houses, office blocks and factories – and they are always in demand. The training takes several years and when they are qualified they can work for a firm or become self-employed.

Plumbers install or replace hot and cold water systems. They fix sink units, toilets, bathroom suites, heating systems, pumps and radiators, and plumb-in washing machines and dishwashers. Some jobs, like putting all the plumbing into a new building, are major. Others might involve a few simple repairs for a householder, such as mending a dripping tap or a leaking radiator.

Plumbers can work for building or plumbing firms, for firms of heating and ventilating engineers, and for major organisations with their own maintenance departments. There are also similar opportunities with the gas boards, oil companies, etc. A popular option is to set up your own business, once you are fully trained and experienced. This obviously means getting some financial backing and advice.

What the work involves

Plumbing is skilled work. Plumbers have to learn to work with a range of materials, including lead, copper, steel, iron and plastics. They need to be able to:

- plan a job;
- follow instructions;
- carry out accurate measurements;
- cut, bend, join and fix pipes;
- use a lot of different tools, like hacksaws, blowlamps, wrenches, drills, soldering iron and spirit levels.

Plumbers use general 'handyperson' skills too. For instance, they remove and replace carpet and floorboards, and fix pipes and radiators where their work will show and must therefore look neat. A basic knowledge of electricity is also needed. Plumbers need to be able to get on with workmates and customers, and to do a clean job, tidying up when they have finished. It's generally necessary to learn to drive, so that you can get from job to job.

Some jobs involve being 'on call' to respond to emergencies with burst, leaking and blocked pipes.

What it takes

Both young men and women can train to be plumbers. As with all jobs in the building industry, you need to be reasonably fit and strong, able to carry heavy equipment, and able to climb and bend. You must be good with your hands and a careful worker. There are no specific academic requirements, but ability in subjects like English, maths, science, design and technology is especially useful.

GETTING STARTED

For school-leavers, the main way into plumbing, as into all building jobs, is through employer-based training. This offers on-the-job experience and attendance at a college or training centre to get National Vocational Qualifications, currently available at levels 2 and 3. Ask your careers adviser about local opportunities. For young people aged 16 to 19 who have at least four GCSEs at grade C, Modern Apprenticeships in plumbing are available, leading to a minimum of NVQ level 3.

ELECTRICAL INSTALLATION & MAINTENANCE

Electricity is vital in our twentieth-century world. It is used for lighting, heating and driving all sorts of machinery. Even systems using other fuels, such as gas-fired central heating, need electricity for their pump operation, timing and temperature control. Telephones, computers and videos would not exist without electricity; farming and food processing industries rely on electricity to feed us. Wherever electricity is used, there is work for installation and maintenance electricians. A Modern Apprenticeship is the main training route and usually requires good GCSEs, including English, maths and science.

Installation

Electrical installation in new houses, factories, offices or schools must be carried out so that electrical equipment can be used safely and efficiently. Electricians must have clear instructions about where cables should be laid, where junction boxes and meters are to go, where switches and power points must be placed. This information is given in the form of detailed plans. Of course, planning for large industrial installations is much more complex than for wiring houses – in domestic settings, electricians may plan their own work from straightforward specifications or drawings.

When a house is being built, the wiring and initial fixing of fittings for electrical power begins before the construction work is finished. This enables cables, protective conduits and switchboxes to be hidden neatly in the walls and under the floors. Electricians work alongside other skilled workers, such as carpenters and plumbers, who do their own share of construction work at the same time. Installation in factories, mines or new ships, say, is far more complicated, but follows the same basic pattern.

Maintenance

Many electricians are involved in the maintenance and repair of a vast range of electrical machinery – from domestic equipment to heavy industrial plant. This type of work may involve dealing with many different machines, or perhaps specialising in one type of machine in a large factory. As well as industrial jobs, electricians can be employed in a variety of settings: hotels, office blocks, central and local government establishments, hospitals, theatres, broadcasting and transport, and the electricity supply companies.

What it takes

To work in electrical installation and maintenance, you need:

- a pleasant manner and good standard of appearance, when the work takes you into the customer's home or workplace;
- to like using your hands and working with tools;
- to be prepared to work very carefully and methodically – electricity is dangerous!
- good colour vision;
- general fitness, to be able to climb ladders, work safely at heights, and get into awkward places;
- to be prepared for unpleasant conditions – it can be very chilly working in a new house with no glass in the window frames and no doors, or outside in the wet and mud.

GETTING STARTED

JTL, the training arm of the Joint Industries Board, offers Modern Apprenticeships in electrical installation. A Modern Apprenticeship usually lasts about four years, and leads to National Vocational Qualifications at level 3. Training is through a combination of learning on-the-job and part-time courses at a college or firm's training school.

You need to find an apprenticeship with a contractor or firm's works department. Your careers service and TEC have details of local opportunities. You could contact JTL directly for help in finding a suitable employer. Most employers will ask for GCSEs at grades A to C, in English, maths and sciences.

Adults can investigate opportunities through JTL or government training schemes. Ask at your local employment services office. Previous relevant experience can be assessed towards gaining NVQs through accreditation of prior learning, or APL.

just THE JOB

HEATING, VENTILATION & AIR-CONDITIONING

> Heating, ventilating and air-conditioning are part of the building services industry, providing for the comfort and well-being of the people who live and work in today's complex buildings. There are jobs at craft level, and for technicians and professional engineers, and the opportunity to work for large contractors, small firms or to be self-employed once you are qualified.

Many modern office blocks are completely air-conditioned, with no opening windows. Industrial air-conditioning is necessary where a dry, dust-free, temperature-controlled atmosphere is vital, for example in electron microscopy suites or high-technology electronics manufacturing bases.

Some contractors specialise in the planning and installation of simple domestic heating systems; others work on major projects such as hospitals, factories and blocks of flats or offices. In a complex project, the design of the heating and ventilation system will be carried out in conjunction with architects and engineers as an integral part of the structure.

There are a variety of jobs ranging from craft-level work, installing pipework and ducting, up to high-level engineering jobs concerned with designing systems for large building complexes. This means that there are opportunities for people qualified from GCSE up to degree level.

As in other jobs in the construction industry, workers may have to work longer hours in the summer than in winter, and may work on unfinished buildings open to the weather.

Jobs in the industry
Fitters and welders
Fitters and welders work on the installation of systems on site. They work from plans and drawings, positioning boilers, pumps

and radiators and fitting pipes. Pipework can include the kind of work also done by plumbers, but larger pipes and ducts will require welding. Fitters/welders use hand tools such as hacksaws and spanners, machines for cutting, threading and bending pipes and welding equipment. The work involves moving around a lot from site to site, sometimes working outdoors and working on ladders and scaffolding. There is some moderately heavy lifting and carrying to be done; hoists are used for the heaviest lifting.

Good fitters might be promoted to charge-hand or foreman/woman.

Mechanics

Mechanics install, maintain and repair equipment such as pumps, boilers and fans. This is again practical work with electrical and mechanical equipment. Work on the maintenance and repair side may mean working on shifts, including nights and weekends, so that essential systems can be repaired immediately. A job with a heating and ventilation contractor will involve some paperwork and a lot of travelling, but you could be more static, working on the plant in a factory or large hotel, for example.

Training will be towards NVQs at level 2, possibly continuing to level 3. Entry for young people is possible through a training placement with a heating and ventilating company. Ask your local careers service, or contact Building Engineering Services Training Ltd as early as possible. Adult entrants may find training is available through Training for Work – ask at the Jobcentre.

Technicians

Technicians can work in a wide range of aspects of the industry, including detailed design and rigorous testing. They are responsible for the drawings – often produced on computers – which the fitters and engineers use to install systems. They handle technical problems which may arise during a contract, organise the supply of materials and coordinate the heating and ventilation work with the work of other trades. With training and experience, a technician could become a **project manager**, supervising the whole process of turning a design into a working system.

Training is through an apprenticeship with attendance at college to take a BTEC National Certificate course in engineering or NVQs at level 3. To be considered for an apprenticeship, applicants need a minimum of four GCSEs at grade C or above, including maths, English language and science.

Modern Apprenticeships leading to NVQ level 3 are available for young people between 16 and 19 years of age.

Engineers

Incorporated engineers do a higher level job than technicians, with a rather different training, giving a broader understanding of the technical background and the industry as a whole. Incorporated engineers may be involved in design and project engineering, testing and commissioning plant, maintenance and service, surveying, estimating, buying and technical sales. A possible entry route is by starting as an apprentice with A level, Advanced GNVQ or BTEC National qualifications, and continuing to study part-time for a BTEC Higher National Certificate. An alternative way is to continue on with full-time study up to BTEC Higher National Diploma level. It can be possible to work your way up from craft level, if you are prepared to study part-time.

Professional engineers – there are good employment prospects for graduate engineers in the design, planning and installation of complex heating and ventilation systems, and also in management. There are specialised degree courses in environmental engineering and building services at a range of universities (see the ECCTIS database or the higher education reference books). Science and engineering graduates can take postgraduate courses in building services. Minimum entry requirements for an environmental engineering or building services degree are usually A level maths and physics, with supporting GCSEs. Advanced GNVQ or BTEC National qualifications may also be acceptable: check with individual institutions.

Adults may be able to find training as a fitter, welder or mechanic through government-sponsored training. Ask at the Jobcentre. Where college courses are concerned you may find that entry requirements can be relaxed in view of your maturity and previous experience. Each case is judged individually.

REFRIGERATION

> Refrigeration is used widely wherever there is a need to store or transport goods which would otherwise perish in the keeping. This applies particularly to the food industry and also to such places as hospitals. There are many other industrial applications, wherever a cooling system is required. It is possible to start in the industry with anything from a few GCSEs to a degree.

To many people, refrigeration means that white box in the corner of the kitchen, which hums away steadily and only gets noticed when it goes wrong. Further afield, you might have noticed the large cold-cabinets in the supermarket or the walk-in cold store at the butcher's. These are, however, only what you could call the tip of the iceberg! Products which end up in the freezer cabinets or cold shelves in the supermarket will have passed through a cold distribution chain from the first chilling or freezing onwards. This chain might have included the factory, warehouses and the transport on the way to the supermarket. A frozen chicken that has defrosted and refrozen may be spoiled and also a health hazard; butter that has become too warm will go rancid.

Refrigeration is vital not only to food supply. There are also, for example, many medical uses for refrigeration. Blood for transfusions must be kept cool, as must many medicines and vaccines. In industry, there is a range of processes which require refrigeration, from cooling rivets which will expand after insertion, to large-scale construction projects where wet, soft ground can be frozen to allow tunnelling or drilling to be carried out. Air-conditioning systems also depend upon refrigeration units.

Refrigeration is one aspect of the building services engineering industry. Jobs in refrigeration can be divided into two main areas:

- **design and installation** of refrigeration plant;
- **repair and maintenance** work, which includes the safe handling and recovery of CFC gases.

Design and installation is done by specialist firms of refrigeration engineers who carry out work under contract. Maintenance and repair work may be carried out by the organisation owning the plant or by a contractor. In the case of a breakdown, repair work must often be done very quickly, so there are firms offering specialist repair services on a 24-hour basis. Weekend and evening work is often necessary.

Levels of work

Though job titles vary, the basic categories are:

Craft worker

This level of job involves mechanical work, assembling and installing systems and carrying out repair and maintenance work. This is practical work with machinery and includes installing pipework and welding. Entrants will usually need GCSEs at grades D/E in mathematics, science and possibly technical subjects. Training is in the workplace and by day release to take National Vocational Qualifications. It usually takes two years to reach NVQ level 2 and another two to reach level 3, although the time taken can vary considerably. Opportunities for school-leavers may be available through training placements with employers: ask at your careers service about possible local opportunities, or contact Building Engineering Services Training (BEST) – see Further Information section.

Technician

Engineering technicians and incorporated engineers are involved with the design, planning and layout of refrigeration plants. They may work on the costing of contracts and purchase of materials and the supervision of work on site. There may also be work in the design and development of new systems and components and in technical sales.

Training involves acquiring a much deeper theoretical knowledge than is needed at craft level. You would usually need four GCSEs at grade C, or equivalent, including mathematics, science and English (or a subject demonstrating the use of English) to get into an engineering technician apprenticeship. You would attend college to follow a BTEC National Certificate course in building services (refrigeration), also leading to NVQs in Refrigeration/Air Conditioning at level 3. Large companies may employ people with an A level in science or maths, plus GCSEs, to train as Incorporated Engineer apprentices to study for a BTEC Higher qualification. Equivalent qualifications such as a relevant Advanced GNVQ may also be acceptable. The careers service and BEST should be able to advise about companies that take trainees.

Modern Apprenticeships in heating, ventilating, air conditioning and refrigeration are now available in many parts of the country for 16 to 19 year olds. A Modern Apprenticeship means full-time employment with a firm which will guarantee training to a minimum of NVQ level 3. Contact your local careers service for further information, or address your enquiries to the Engineering Services Training Trust Ltd (see Further Information section).

Professional/chartered engineer

These are the graduate-level jobs in the industry. At this level, engineers carry out design and development work, supervise contracts and in general do the work on complex installations which requires the depth of knowledge of the graduate engineer. Management posts are usually filled by staff who have entered the industry through this route. There are no first degree courses specifically in refrigeration, but building services engineering courses include heating, ventilating, air conditioning and refrigeration modules. A few colleges and universities offer HNDs in engineering which include a refrigeration option. There are one or two postgraduate courses in refrigeration.

Minimum qualifications for entry to engineering degree courses are usually A level mathematics and physics, with supporting GCSEs. Universities would normally look for three A levels. A relevant Advanced GNVQ or BTEC National qualification is also usually acceptable.

COULSDON HIGH SCHOOL
HOMEFIELD ROAD
OLD COULSDON
SURREY CR5 1ES
TEL: 01737 551161

just THE JOB

STEEPLEJACK

Steeplejacks work on tall structures like chimneys, monuments and church spires, using ladders, rope and scaffolding. They repair and replace damaged brickwork, stonework and tiles, fit lightning conductors, replace glass, weld and paint. They may also take photographs and write reports on the condition of a structure.

What it takes

- Obviously, a head for heights is a basic requirement.
- Steeplejacks must also be fit, sure-footed and able to climb and balance.
- Not only do they have to climb heights, but they have to do a job of work when they get to the top, so they need good practical skills.
- Most of the work is done out of doors, and steeplejacks must be able to cope with all weather conditions.
- They must be careful, conscientious workers. Their own safety, and that of other people, depends on their skills.

Employers would expect trainees to be of at least average general ability. Subjects like science, maths, design and technology are the most useful ones.

GETTING STARTED

The steeplejack industry has set up a special scholarship-apprenticeship scheme with the Construction Industry Training

Board's Civil Engineering College at Bircham Newton, Norfolk, for a small number of school-leavers (12 places) wishing to take up apprenticeships to become steeplejacks. This is a two-year programme for 16 to 22 year olds, working towards National Vocational Qualifications at levels 2 and 3. It provides 24 weeks' residential training at the College in the first year, and four weeks in the second year. Trainees also spend two weeks on things like Outward Bound activities, which give experience of leadership and working in a team. When not at the College, the trainees gain practical work experience with their sponsoring company. This programme acts as the entry point to the industry's Apprenticeship Scheme for Steeplejacks.

There is also a Lightning Conductor Engineering Apprentice training programme, run along similar lines, and working towards NVQs at levels 2 and 3. It involves fourteen weeks' off-the-job training at Bircham Newton in the first year and eleven weeks in the second year. This programme calls for slightly higher academic qualifications (GCSEs at grade E and above) as there is some work with electrics. As with steeplejacks, apprentices for this training programme are sponsored by a company, and have employed status from the start. The apprenticeship is open to 16 to 22 year olds.

Other young people and adults can train as steeplejacks 'on-the-job', with the possibility of attending training modules at the College at Bircham Newton.

STONE CLEANER

> Stone cleaners brighten up our lives! In cities and towns, dirt and fumes from constant heavy traffic blacken the buildings and monuments. Stone cleaners work to get them looking as good as new.

Stone cleaning is done with sandblasting equipment and wire brushes. The stonework is washed down with a caustic solution, or sprayed with high-pressure steam or water. It may also be part of the cleaner's job to repair damage, fix new masonry, etc, though this might also be done by a stonemason.

What it takes

To be a stone cleaner, you need:

- to be fit and healthy, and suited to outdoor, practical work;
- a head for heights and a good sense of balance, to cope with ladders and scaffolding;
- to be able to follow safety regulations and wear protective clothing;
- to be prepared to work in dust and damp conditions, and to put up with the noise of equipment.

Qualifications aren't required, but they always help. Employers will be looking for careful and responsible workers.

You'll travel around

As a stone cleaner, you work in different places all the time, within your firm's work area. This variety appeals to some people. The day-to-day work, however, remains the same.

Stone cleaners work for private companies and are trained on-the-job, possibly leading to an NVQ level 2 in facade maintenance (another way of saying stone cleaning). You could contact firms about possible job vacancies.

If you are a school-leaver, you could ask the careers service whether there are training opportunities available which will provide you with skills and experience. Adults might be able to find a training opening through government-funded schemes; the Jobcentre will have information.

just THE JOB

OPERATING & MAINTAINING CRANES & HEAVY PLANT

> In the construction and civil engineering industry, many different machines, like cranes, diggers and tractors are used. Operators are needed to work them, and mechanics are needed to service and repair them. All these heavy machines are expensive, and can be very dangerous if carelessly handled. So operators and mechanics must be well-trained and responsible.

Operating the different types of construction machinery is a skilled task. Most operators specialise in a particular kind of equipment, such as earth-moving machines, mechanical diggers, rough terrain fork-lift trucks, dump trucks or cranes.

Plant mechanics have to ensure that all the heavy plant and machinery is working properly. Machinery that isn't working costs a contractor time and money, and causes delays for other workers on the site. The work is varied: you may be in a workshop one day working on an engine part and out on site the next, fixing the gearbox of a tower crane.

What it takes
- You usually need an ordinary driving licence, with some driving experience.
- It will help if you know something about petrol and diesel engines.

- Employers prefer applicants who are over 18, and probably older.
- You need to be a careful, reliable worker.
- You have to put up with dust, dirt and noise.
- Being fit is important; if you suffer from asthma, epilepsy or migraine you would probably find the work unsuitable.
- You need a good head for heights.
- Working outside in cold weather is part of the work, although many vehicles have closed cabs.

Safety is important!
Everyone in the construction industry must think about safety

when they are working and use their common sense. Serious accidents can easily happen, and all workers need to be able to trust their mates to work safely.

GETTING STARTED

Training is generally given on-the-job. Some firms send their operators on a short course at the Civil Engineering College at Bircham Newton, Norfolk. This makes sure you learn to operate machinery safely and correctly.

Plant operators are often chosen from workers already employed by the firm, so any job with a construction, civil engineering or plant hire company would be a good start. Look in your local *Yellow Pages* telephone directory for addresses. Water companies, local authorities and the quarrying industry also employ plant operators and mechanics.

There are a limited number of places on the Construction Industry Training Board's Construction Plant Technician and Plant Mechanic Training schemes. These take up to three years, and involve work experience plus time at Bircham Newton. They lead to National Vocational Qualifications up to level 3 for mechanics, and BTEC National Certificate in Civil Engineering for technicians. Plant operatives who undertake and complete training successfully can be issued with a CTA (card of training achievement), which is like an identity card which lists their skills. Ask your local CITB office for details of all these schemes. You may also be interested in training opportunities in other aspects of building, offered through the CITB.

School-leavers wanting to get into a construction or civil engineering firm should contact their careers service for information on local training opportunities. Adult jobseekers should contact their local Jobcentre.

just THE JOB

DEMOLITION

> Demolition workers clear sites of old buildings and unwanted rubble before any new building work or landscaping is done. They also have to knock down buildings which have become unsafe, because of age or damage resulting from fire, gas or other explosion.

What the work involves

The exact job to be done is a bit different for each building which has to be demolished. A foreman or supervisor has to assess each job to see just what needs doing, and in what order. Demolition work has to be done with a great deal of planning and care. It is easy for serious accidents to happen if thorough preparation has not taken place.

Parts of a building may need to be reclaimed, so that material can be sold for re-use. So the demolition workers may strip roofs and floors, and remove fittings such as old fireplaces, leaving just the skeleton of the building.

Various ways are used to knock down the walls of the building. A huge ball and chain may be swung at them to demolish them, and equipment such as bulldozers, air-powered chisels and pickaxes are then used to further break up the structures. Some demolition workers will train to operate bulldozers, cranes, grappling equipment and other specialist machinery. Some will learn to use explosives, but this is specialist work. Other specialist jobs, for which you need extra training, include using oxy-acetylene cutting equipment and working with dangerous substances like asbestos.

With training and experience, you may be able to progress to foreman or forewoman or to supervisory jobs. The traditional titles of *'mattockman'* and *'topman'* are still used for such jobs.

What it takes

- Demolition work is a very physical job, so strength, fitness and stamina are essential.
- Qualifications are not required to get started, although maths and English are important if you progress to a position where you need to measure, calculate, and interpret plans and instructions.
- You need commonsense and an ability to work carefully and methodically.
- You would almost certainly need to be over 18, and a general background of building site work is useful.
- You need to be prepared to work in dirty, dusty and noisy conditions, and in unpleasant weather.

Safety is important!
You have to be very safety-conscious (helmets, goggles, etc, are worn) and you have to follow instructions to the letter. There is always a risk of falling or being injured. Demolition workers operate in a gang, and each member of the team is responsible not only for his or her own safety, but that of everyone else too.

You could work anywhere!
Demolition gangs usually do quite a lot of travelling about, working for a few days here and a few days there, wherever there is work to be done. There may also be shift or weekend working. These conditions can disrupt your family or social life, but, because of the dangerous nature of the job, pay can be high.

GETTING STARTED

A school-leaver hoping to work in demolition might get started by finding a job or training place connected with the building industry. A training place as a building operative or steeplejack, for instance, would give you up to two years' relevant on-the-job experience, with the chance to attend a college or training centre to gain recognised qualifications. The Construction Industry Training Board offers training towards NVQ levels 2 and 3 in Demolition.

Adults would usually be expected to have had some relevant experience in the construction or civil engineering industries, though there are jobs where 'muscle' is the main requirement.

Once you are working in demolition, your employer might enrol you on the Scheme for the Certificate of Competence of Demolition Operatives provided by the CITB. This involves both on-the-job and off-the-job training.

just THE JOB

BUILDING TECHNICIANS, TECHNOLOGISTS & MANAGERS

> There are many jobs in the building industry which involve organising, financial management, technical knowledge, or negotiating and planning skills, rather than manual work. Some jobs are on construction sites, while others may be office-based. There are opportunities ranging from those which require some good GCSEs to degree level.

To understand the range of jobs, it is necessary to look at what is involved from start to finish in putting up a big structure, such as a supermarket or bridge.

Design
The first stage in a building project involves architects and civil or structural engineers. If the design is for a major building or construction, other specialists may be involved, such as environmental engineers, landscape architects and building services engineers. When the project has been designed on paper, a quantity surveyor works on it to produce a bill of quantities – literally, a list detailing everything which will be needed to turn the design into reality.

Tender
The design and bill of quantities are sent out to a number of contractors (building firms) who are asked to submit a tender. The tender is the price for which the contractor says the job can

be done. At the contractors, estimators, buyers, planners and site engineers work on the tender. The architect or engineer who is working on the client's behalf recommends which tender to accept, and the contract is awarded.

Preparation/planning

The next stage is a detailed plan of how the work will be tackled. Materials are ordered, detailed site inspections and measurements are made, and specialist workers and equipment are hired or booked.

Construction

Managing the construction work is a big job. Essential work includes progress chasing (*why didn't those bricks arrive on time?*) and checking and measurement as the work progresses. Bad weather may delay construction, and the design may be altered as work goes on.

Each of these stages in a construction project involves work for various professionals, technologists and technicians.

Professional/graduate work in building firms

There are professional and managerial opportunities in the building industry for people with qualifications such as a degree or Higher National Diploma in building, or membership of the Chartered Institute of Building. In large firms, these jobs will be specialist posts; in smaller firms, more than one function will be done by one individual. For instance, site management and site engineering may be combined, or estimating, quantity surveying and planning. As with technician-level work, some jobs are largely office-based, some largely site-based, and the majority are probably a mixture.

Technician-level work

Technicians work on detailed technical parts of a project, rather than overall planning and supervision. They do a variety of jobs in all parts of a project, including planning, estimating, surveying and project management. They may train for one particular job, but there is quite a lot of movement between specialisms, as people's careers progress. In smaller firms, staff have wider responsibilities and will do more than one of the jobs described.

Some of the areas of work involved

Site or project management

The site or project manager (sometimes called the *site agent*) is the person in overall charge of a building project. He or she makes sure the project is running smoothly and profitably. On a big project, there will be a large number of building workers, including many subcontractors. The site manager must be a very good organiser, with an all-round knowledge of building and the ability to get on with, and manage, people.

Estimating

Estimators and estimating technicians do work such as tendering for contracts. When someone wants a building job done, they will ask a number of companies for quotes. Building firms need staff who can work out how long a job will take, all the materials which will be needed, and which suppliers and subcontractors will offer good prices and reliability. Estimators need a good knowledge of building construction. They must be good at calculations and able to work under pressure.

Buying

Buyers or purchasing officers have to purchase the materials needed for a job. They have to know their suppliers well, and get the best prices and delivery arrangements that they can. This is basically an office job, with a lot of the negotiating done over the phone. Buyers have to ensure that the suppliers can deliver materials on time, because delays in deliveries can mean big problems with a job. Workers and machinery may be left idle, and the whole contract can be in trouble.

Planning

Planners are responsible for deciding the order in which a particular building contract is carried out. They need an eye for detail and knowledge of construction methods. They work out the most economical way of using all the resources it will take to do the job: building craftsmen and women, machinery, equipment and materials. The planner sets targets for the completion of certain stages of the job, and monitors its progress once it is under way.

Site engineering

With the assistance of technicians, the site engineer surveys the construction site before work begins. Measuring and surveying instruments are used and soil samples may be sent for analysis. The engineer makes sure the site is marked out exactly as the designer has specified. When the building work is in progress, the engineer has to check that the work is being done accurately and the right materials are being used.

Quantity surveying

Contractors' quantity surveyors (or cost and bonus surveyors) are employed to work out the amount and value of work actually done by the people working on a construction job. This involves visits to the building site. They prepare intermediate and final accounts for the client and get involved in financial and legal issues relating to the building contract. If alterations to an original design occur during construction, they calculate the cost and come to an agreement with the client's quantity surveyor. It's a job involving work with figures and a methodical approach, as well as the ability to deal and negotiate with people. (See next section.)

Clerk of works

A very modern job with an ancient title: the first clerks of works were monks who supervised the building of monasteries.

The modern clerk of works is employed by the owner of a building under construction to check that the contractors are carrying out the work correctly as specified in the contract. They need expert knowledge of all aspects of construction and can specialise in particular areas such as heating and ventilation and electrical installation. Historic buildings such as cathedrals and stately homes also have clerks of works, who are responsible for the maintenance and repair of the buildings in their care.

Qualifications can be obtained initially through BTEC courses followed by the specialised exams of the Institute of Clerks of Works (see Further Information section).

Building maintenance management
This is a specialist area of the construction industry and involves looking after the building once the construction process is completed. It is a very broad area: some maintenance personnel will deal only with building services such as heating and ventilation systems, whilst others may deal only with the fabric of buildings, i.e. walls, roofs, floors and foundations.

Other professional/graduate/technician work in the construction industry

There are many other opportunities for people with degrees and equivalent qualifications. You might be interested in working as an architect, chartered surveyor, client's quantity surveyor, building surveyor, consulting engineer, civil or structural engineer, building services engineer, landscape architect or town planner.

All these professions also need technicians to support them, with qualifications similar to those listed below.

TRAINING

Technician training normally requires a minimum of three

GCSEs at grade C, strongly preferred subjects being English, maths and science. Many FE colleges and sixth forms offer an Advanced GNVQ in Construction and the Built Environment or BTEC National Diploma courses in construction or building studies. It is also possible to train whilst in employment, studying for a BTEC National Certificate. This generally lasts up to two years, involving both on-the-job experience and attendance at college.

Modern Apprenticeships organised through the Construction Industry Training Board (CITB) are available and can lead to technician-level work. The electrical training organisation, JTL, offers electrical technician training. Find out about training opportunities in your area through your local careers service.

From technician to professional qualifications
Following an Advanced GNVQ, National Certificate or a National Diploma, you can progress to a BTEC Higher National Certificate or Diploma. It would take two years of part-time work to gain a Certificate and three years part-time, or two years full-time, to reach Diploma level. Alternatively, you could take a degree or professional course. You can also work for other professional or 'higher technician' qualifications by part-time study, such as the examinations of the Chartered Institute of Building or Royal Institution of Chartered Surveyors.

Direct entry to professional training
You can take a degree or BTEC Higher National Diploma course in building. Degrees require at least two A levels, a GNVQ Advanced or a BTEC National Diploma/Certificate; HNDs require at least one A level. Supporting GCSE subjects should include maths and science, and design and technology is also relevant.

Both HND and honours degree courses are available at many universities and institutes of higher education, on a full-time or sandwich basis. You can find details of courses in the *Official UCAS Guide to University and College Entrance* or ECCTIS database. Some national contractors offer sponsorship to under-graduates at universities.

QUANTITY SURVEYOR

> Quantity surveyors are experts in the cost and management of construction projects. They may work in private practice, for a building contractor or for commercial and industrial companies. Most quantity surveyors have degrees, although there are also opportunities at technician level.

Quantity surveyors act as a link between the architect and engineers on one side, and the contractor on the other. They are key people, involved at all stages of a project. A major part of the cost of any building or civil engineering project is represented by the materials to be used. How much steel or concrete is needed to build a motorway flyover? What price can a particular type of steel be obtained for? Who can supply it? In calculating the cost of a project, many factors of this sort must be taken into account.

What the work involves

A quantity surveyor . . .

- works with a contractor to produce initial plans and drawings for a project, and helps to calculate the most economical way to do the job;
- assesses the 'quantity' of labour, materials and equipment required to complete the project;
- prepares a 'Bill of Quantities' from which an estimation of the entire cost of the project can be drawn up by the contractor;
- assesses interim payments to be made for materials;
- follows the progress of the project to make sure it is completed on time and within the budget.

What it takes

A quantity surveyor needs . . .

- a good head for figures;
- a sound knowledge of construction;
- knowledge of industrial and contractual law;
- organisational skills;
- communication skills.

Steve – quantity surveyor

'My job is largely about getting good value for money. I work from the architect's plans, calculating how many bricks, tiles, sacks of cement and so on are going to be needed for a job. I need to know about the uses of different materials, where to buy the best quality at the best price, and so on. I also have to predict how many craftspeople will be needed for how long, and then work out how much the whole job is likely to cost from start to finish. I work with the architects, the builders' merchants and the builders themselves, so I get to know a lot of people in the industry. It's a very responsible job, because if I get my sums wrong, it can mean the difference between profit and loss!

The work is a mixture of desk duties and site visits, which means that every day is different. I enjoy the hard hat and wellie days, but sometimes it's nice to be able to retreat to a cosy office.

To qualify, I first had to get good GCSEs and then two A levels (an Advanced GNVQ in Construction and the Built Environment might be an alternative to go for now – it wasn't around when I was at school) before going to university to study for three years. I became a student member of the Royal Institute of Chartered Surveyors while on my degree course, and after graduating and working for a year was able to qualify as a chartered surveyor.

I'm now qualified to work anywhere in the European Union. Maybe, one of these days, I'll get a job where they're building a luxury holiday villa complex somewhere on the Mediterranean!'

Where do quantity surveyors work?

Quantity surveyors are employed in large contracting and civil engineering firms, in local and central government departments and in private practice, such as specialist firms of quantity surveying consultants. More than half of all quantity surveyors work in private practice. People who like the outdoor aspect of the work will find a job with a contractor the best alternative.

TRAINING

In order to qualify as a **professional surveyor**, you must successfully complete an accredited degree or diploma. This is followed by a two-year period of practical training whilst working, and a final examination, by the professional body of which you are seeking membership, to test your competence. To start a degree or diploma course, you need five GCSEs at grade C, including maths and English, and two A levels. An Advanced GNVQ/BTEC National Diploma or Certificate may also be acceptable. Check with the appropriate universities for the entrance qualification required. A relevant BTEC HND/HNC may gain entry directly to the second year of a degree or diploma course.

Technical surveyors are also employed on all aspects of the work, to carry out the functions which do not require the expertise of a fully qualified surveyor. To become a technical surveyor, you need four GCSEs at grade C, including mathematics and English. You can train either whilst you are in employment, or full-time at an FE college. Further details can be obtained from the Society of Surveying Technicians (see Further Information section).

Professional examinations in quantity surveying are also offered by the Royal Society of Chartered Surveyors, the Architects and Surveyors Institute, and the Chartered Institute of Building.

ARCHITECTURE

> An architect designs buildings of every kind. To become an architect, you train over a period of years to gain the skills needed for creating attractive, safe and usable constructions that will stand the test of time! With four GCSEs at grade C, you can usually start training for an Advanced GNVQ in the Built Environment and, with this equivalent to two A levels, continue training with a Higher National Diploma course in Building Studies or a degree course in Architectural Technology. Realistically, to be accepted onto a degree course in Architecture you will probably need to have a minimum of three good A level passes, or their equivalent.

Architect

The architect's job is complex and technical, and architectural work as a whole involves a lot more than sitting at a drawing board or a computer screen, giving expression to creative ideas. Architecture is closely connected with people: a large part of the work involves talking to clients, planners and committees about designs, and the purposes for which buildings are needed. Architects visit building sites frequently to see how their design is being constructed. Patience is a virtue in this work; some projects take many years to reach completion.

All that you need to know can't be learned in a few weeks or, indeed, in a few years. An architect's training takes seven years and, at the end of it, you still do not have enough experience to run a major project. If you are setting out to become an architect, it's going to be a long journey.

What the work involves
Functional design
Buildings have not only got to *look* good: their design has to closely match their *function*. Consider part of the range of buildings which you could be involved in designing, such as:

- stores, factories and warehouses;
- sports complexes;
- music centres and concert halls;
- schools and libraries;
- crematoria;
- health clinics and hospitals.

There are always specific requirements for every building, such as provision for disabled people using the building. You also have to plan for things like fire escape routes, adequate ventilation and access for emergency vehicles.

Technical knowledge
Part of the architect's job is to ensure that a building is constructed properly and that it stays up! The quickest and cheapest way is not always the safest or the most long-lasting – so technical knowledge is important. You must know about the properties of different building materials to understand both their advantages and their drawbacks. You also have to know about things like plumbing and drainage, landscaping, soil mechanics and building law.

Budgeting
Of course, the money available for a project is always the first and the biggest consideration. What can you achieve within the available budget? This is where there is a lot of compromise between the ideal and the affordable, and it's up to the architect

to point out the implications of penny-pinching on the projected look and functioning of the completed building.

Where do architects work?

Almost a third of qualified architects are situated in and around the London area, but, at the present time, architects are underemployed. The construction industry is always very sensitive to booms and slumps in the economy, which affect demand for architects in both the private and the public sector. Unfortunately, an increasing number of large building firms develop housing estates to a well-tried formula, cutting out the architect almost completely.

Architects can work either in public service, for local councils and health authorities or trusts, or in private practice. Private practice can range from a one-person practice to the enormous firms which build estates and factories all over the country. A typical firm has about twenty staff. Architects may be taken on as salaried assistants and, with experience, become partners.

In public service, county, district and borough councils are major employers of architects, although job opportunities have shrunk considerably since local government reorganisation. Local government architects design buildings, and also oversee the design and construction of buildings being put up by contractors. Earnings are not as high as the potential fees of private practice architects, but a dependable salary can be more attractive during an economic recession, when there is always a slump in the construction industry. Various Civil Service departments also employ architects.

Don't forget that British architects' qualifications are recognised and well-regarded within the European Community, and many British architects find work overseas.

Laura – architect

Laura is one of two women architects in a medium-sized practice which handles quite a range of work. She has welcomed the introduction of CAD (computer-aided design) into architecture. 'It has really speeded up the stage of producing detailed drawings,' she says, 'but some of the older partners and technicians are not interested in adopting a new technology.'

Laura enjoys the challenge of producing excellent designs, given all the constraints under which architects have to work. 'Money is always the biggest issue. The client may not be able to afford the materials and interior fittings which would really do your design justice. You have to compromise. Then there are the limitations imposed by the actual site, planning restrictions and building legislation, safety requirements, the need to provide access for disabled people if it's a public building, and so on.' Laura emphasises the technical side of the job: 'Although you have to have artistic flair, it's no good just being arty – you really have to know how structures work, and you mustn't mind going out on a muddy site.'

Teamwork and management skills are also vital. 'Running a project means coordinating the work of everyone involved, and you have to make a lot of site visits to check that the job is progressing properly.' Working hours are typically 9.00am to 5.30pm, but the day can last much longer if the pressure is on to meet deadlines. And, in a working day, Laura often has to move between mucky building sites and office-based meetings with clients. 'I always have a pair of smart shoes in the car, as I can hardly go from a building site to a meeting with my next client wearing wellingtons caked in mud!'

Specialisms
Architects may decide to specialise in a particular aspect of work; this possibility can arise in both public and private practice. You can follow specialist postgraduate courses in fields such as the conservation and restoration of historic buildings, garden and landscape architecture, interior design, landscape design, management, research and teaching.

EDUCATION AND TRAINING

You can't practise as an architect unless you are registered with the Architects Registration Council of the United Kingdom (ARCUK). The normal route to qualification as an architect involves attending a school of architecture recognised by ARCUK and the RIBA (Royal Institute of British Architects) on a three-year, full-time degree course. This is then followed by a two-year Diploma/Bachelor of Architecture course. In addition to these five years of academic work, training includes two years' practical experience. The first year of architectural practice is usually completed between the degree course and starting work for the diploma, the second year always after the diploma course. The professional practice examination is then taken, leading to mandatory registration, and membership of RIBA if desired. Full-time (and some part-time) courses are available at 36 recognised schools of architecture attached to universities and other institutions. A list of the Schools of Architecture, with details of courses recognised by the RIBA, is published by the Royal Institute of British Architects.

RIBA offers guidelines to the recognised schools of architecture on minimum educational requirements, which are quite strict, so read this section carefully. You need a minimum of two A levels, or one A level plus two AS levels, supported by at least five subjects at GCSE grade C, including English, maths and double science. Both A levels and two of the other GCSEs must be in subjects regarded as 'academic', such as science, history or modern languages. Freehand drawing ability is necessary, but you don't have to have exam passes in art. However, you will almost certainly need to assemble a portfolio to show at interview. The RIBA recommends that a mix of both arts and science subjects should be studied. Many schools of architecture have higher requirements than the recommended minimum qualifications (some ask for three A levels, including mathematics or science), so it is always sensible to check with the schools in which you are interested.

Most schools of architecture accept an Advanced GNVQ in Construction and the Built Environment, with modules passed at distinction, as fulfilling their entry requirements, sometimes with an additional A level.

Students with the BTEC National Diploma or Certificate in Building Studies are advised to contact schools of architecture in order to check whether the qualification will fulfil specific entry requirements.

Architectural technicians and technologists

Technicians do not need to have the full range of professional architectural skills, but the work of an architectural technician is still technically demanding. Technicians are employed in both private and public service work in the same way as architects. Like architects, they now use computer-aided design as an important tool. Some job advertisements ask for architectural technologists. These positions carry more responsibility than those of technicians.

What the work involves

The tasks undertaken by a technician can vary considerably from one practice to another. They may include any or all of the following:

- investigating, analysing and preparing technical information needed for the design of a building;
- preparing the drawings and schedules which builders use to construct the building, often using CAD techniques;
- doing presentation drawings, charts and diagrams for clients;
- dealing with the administration of contracts;
- liaising with specialist advisers;

- land and building survey work;
- site meetings;
- inspections of work;
- collecting information on the long-term performance of finished buildings;
- office management.

Sayeed – architectural technician

I work in a practice which has three architects who run the business as partners. Then there's the senior technologist, myself, a secretary and a part-time clerk.

I would have liked to be an architect too, but the training is so long – seven years altogether – that I decided I wouldn't last the duration! I got myself taken on as a trainee after getting five good GCSEs, including maths and English, and studied at college on day release. Being a technician is really quite challenging and interesting and the more the architects learn to trust you, the more responsible the work they give you to do. A lot of my job is producing three-dimensional drawings on computer – all very high-tech.

An experienced technologist or technician can even go into partnership with an architect to build up a practice. Obviously the architect is the one whose design ability and professional training gets the contracts, but the technician shares a lot of the technical knowledge about building techniques and materials, planning law and so on, so can play an important part in the business.

I still get quite a buzz from seeing one of our contracts completed. It's exciting, watching something which was just a few lines on a sheet of paper become a real building.

EDUCATION AND TRAINING

Office-based training

The economic recession has meant a decline in vacancies for trainee technicians. Technicians can start by joining an architect's office at age 16–17 with about four GCSEs at grade C, or their equivalent. Subjects must include mathematics, science and a qualification which demonstrates an ability to write English. You don't need to have studied design technology at school, though it could give you a useful background to design work and problem solving. You need technical and drawing skills, together with the ability to work accurately, paying close attention to detail. It is very important to be able to work as part of a team.

Training is by day release to college to study for a BTEC National Certificate in Building Studies. Although GNVQ courses have been introduced all over the country, there are still colleges which offer the BTEC National courses, particularly for part-time students. Check this with the colleges in your locality. Subjects studied during training include construction technology, materials, structures, surveying and levelling, building science and mathematics. You can later go on to take the Higher National Certificate.

Full-time/sandwich courses

An alternative route is to take a two-year full-time course leading to the BTEC National Diploma in Building Studies, for which the entry requirements are the same as for the Certificate. For students with one A level plus four GCSEs at grade C (subjects to include mathematics, science and a subject demonstrating use of English), or a BTEC National Certificate/Diploma in Building Studies, there are Higher National Diploma courses in Building, specialising in architectural technician work. These courses last two years full-time, or three years part-time.

The Advanced GNVQ in Construction and the Built Environment (which has largely replaced the BTEC National Diploma) is also acceptable as an entry requirement for Higher National courses; check with individual institutions.

Membership of the British Institute of Architectural Technologists is open to those with Higher National qualifications or a degree, after gaining work experience and compiling a logbook. Some candidates may be exempt from the academic qualification, subject to their work experience and ability. People still studying for the qualifications can become student members of the Institute.

Can a technician become a professional architect?

It is possible for a technician to qualify as an architect by further part-time study over three to five years. There are one or two modular degrees which may exempt technicians from some modules, but this is unusual. A technician with suitable qualifications is more likely to qualify by taking a full-time degree course in the usual way. But remember that the two jobs are rather different and architects need to be more creative, with a flair for design.

Adults: because of the long training, architecture is not an ideal second career. Check with individual institutions whether or not they accept mature students.

CIVIL & STRUCTURAL ENGINEERING

> Civil and structural engineering is about the planning, design and construction of large systems and structures. This means things like roads, bridges, dams, water and drainage systems, energy supply systems and large buildings. Civil and structural engineers spend some of their time in the office and the rest out on site. Engineers have degrees or equivalent qualifications, but there are also opportunities to train on-the-job as craftsmen/women and operatives.

A lot of knowledge and many different skills are used in civil and structural engineering. It's very much a team effort. To look at this further, think about all the processes which go towards building something like a road bridge.

Getting the job done

Once it has been decided where to site the bridge (and this is a complicated decision, with environmental factors, transport, and community needs to be taken into consideration), **consulting civil engineers** must decide which type of bridge is going to be most suitable. This depends on the loads it must carry, the type of foundation available and the span to be crossed, what materials are to be used, how it will fit into its environment, and how long it will take to build.

There may be several possible solutions to a particular problem and the **structural engineer** has to balance the need for strength against the cost of construction. It might well be possible to construct a bridge to last 2000 years, withstand a 200 mph gale and support two lanes in each direction, solid with 40-tonne lorries. This would be so expensive that no one would want to pay for it. However, there are certain stresses which a bridge *must* be able to withstand.

Highway engineers may also be involved in the design of the bridge – especially the road surface itself and its place in the overall road system in the area. Should it be motorway standard, or an ordinary two-way road? Can the bridge itself or the connecting roads be cheaper or more environmentally friendly? What about safety and future maintenance? Gradients and earth-moving, specifications for foundations and surfacing are all problems for which the highway engineers and technicians must find solutions. Two hundred years ago, Britain's roads were the worst in Europe. A lot has changed since then. Roads are now the most heavily used transport system we have, leaving railways and canals far behind in importance.

Once a road is built, the highway engineer's job is not over. Maintenance, as anyone who has ever driven up the M1 will know, is a continuous process. So many engineers and technicians are involved in surveying, checking and repairing the existing road network, and updating estimates of the volume of traffic served.

When all the discussions are over and the decisions made, **civil engineering contractors** are invited to offer to do the job, quoting a price. Several contractors will quote, and one will be chosen. The successful firm will have to make the bridge for that price, in order to make a profit. Mistakes at this stage will mean a loss (potentially catastrophic on a multi-million pound project). Then construction work can start on site, involving a large team of engineers, technicians and operatives.

A large project, such as a major suspension bridge, will take several years to complete. A bridge is a comparatively straightforward example of a civil engineering project; imagine the complexities of a huge scheme like the Channel Tunnel or the Thames Barrier.

Who works in civil and structural engineering?
- **Craftsmen/women and operatives** do the skilled manual work and machine operating;

- **Incorporated engineers and engineering technicians** work on the detail of projects, such as costing, site management, testing, computerised drafting and surveying;
- **Graduate engineers** design and manage civil engineering projects, act as consultants, etc, and can become chartered engineers after training.

Craft workers and operatives

There are a variety of practical jobs for which academic qualifications are not essential, though qualifications are always helpful. Training could take several years. Many skilled craftworkers are self-employed. They also carry high responsibility for safety on site.

Jobs on site include concreting, scaffolding, steelwork and operating bulldozers, cranes, diggers and other machinery.

Incorporated engineers and engineering technicians

Both incorporated engineers and engineering technicians work on the detailed aspects of designs, producing working drawings, doing calculations, costings and progress chasing. Much of the work is done on computer. They would be responsible for solving many of the day-to-day engineering problems that occur during any large project. Tasks they get involved in would include analysing and testing samples of materials, and work on site — surveying and supervising construction, for example. There is scope to specialise.

Incorporated engineers often take on managerial duties, acting as team leaders to engineering technicians, and they are responsible for the efficiency of the team.

Employment openings at technician and incorporated engineer level include jobs with consulting engineers and contractors, local authorities and other public service organisations, utilities and firms involved in transport and energy supply. Like graduates, engineering technicians and incorporated engineers can also go abroad for work, where their skills will be much in demand.

Chartered engineers

Graduate civil, structural and highway engineers have the overall responsibility for design and research work and for site operations. They may work for firms of contractors, consulting engineers, local authorities and other government bodies, and organisations involved in traffic, transport and energy. Graduate engineers entering at this level would usually aim to achieve professional qualifications, in order to become chartered engineers.

Civil and structural engineers do a mixture of office-based and site-based work, often with building or repair and maintenance contractors. Structural engineers may assist architects in the design of large buildings and other free-standing structures.

Municipal engineers, **highway engineers** and **water engineers** often work for public service organisations like local authorities, the Department of Transport and water companies.

Consulting engineers spend most of their time on creative design and specification work, working to the client's instructions. Engineers employed by contractors see projects through from design stage to construction, including project management.

There are good employment opportunities overseas, both on highly paid contract work (e.g. for foreign governments), and in developing countries, through voluntary work organisations such as VSO.

TRAINING AND QUALIFICATIONS

Craftsmen/women and operatives
For this sort of work, you normally train on-the-job or through a Construction Industry Training Board (CITB) training scheme. People aged 18 or over are generally required for operating construction plant or where there is a need to be able to drive. For younger entrants, apprenticeship training is organised by the CITB. While training you would work towards National Vocational Qualifications. Ask your careers service for details about training opportunities in your area.

Engineering technicians and incorporated engineers
You can start training for technician level after taking GCSEs, as a trainee in a civil/structural engineering design office or with a contractor or local authority. You would then study part-time; currently this would be for the BTEC National Certificate in Civil Engineering. Trainee technicians normally need four GCSEs at grade C, including maths and science. It is sometimes possible to start with lower qualifications.

An alternative is to carry on with full-time education after GCSEs, taking an appropriate Advanced GNVQ or BTEC National Diploma. Having reached this level, to become a registered technician (Eng Tech) you would need to gain some experience in the workplace and pass a professional review set by the professional body, such as the Institution of Civil Engineers, or one of the other institutes listed in the Further Information section.

Technicians or holders of BTEC National/Advanced GNVQ can study part-time, or continue with a full-time course at university or college, to gain a BTEC Higher National Diploma or Certificate. You should check that your course is one which is approved by the relevant professional body. Having reached BTEC Higher qualifications and gained about two years' experience in the workplace, you would need to pass the professional review set by the professional body in order to become registered as an **incorporated engineer** with the Engineering Council.

At whatever level you enter the industry, there should be opportunities to continue with part-time study and training. It is possible for technicians to work up to incorporated engineer level, and for incorporated engineers to reach graduate and eventually chartered status, by taking a degree on a part-time or full-time basis. It is also possible to work your way up via the NVQ route.

Chartered engineer
The first step to chartered engineer status is an honours degree. Usually this will be in civil or structural engineering, though graduates in related subjects (other branches of physical science, engineering, etc) may also move into civil engineering.

Minimum qualifications for an appropriate degree are two (preferably three) A levels with supporting GCSEs. At GCSE level, maths and science are essential. English, modern languages, and design and technology are also useful. A level subjects should include mathematics and preferably physics or engineering science. If you do not have science A levels, there are a number of engineering foundation courses available, linked to various engineering degrees, which you follow for one year before moving on to the first year of the degree programme.

An Advanced GNVQ in a relevant subject is an alternative entry qualification to A level. Check entry requirements with individual institutions, as some may require you to have taken additional GNVQ units, for example. BTEC National Diplomas and Certificates are also alternative qualifications.

Degree courses usually last three years (full-time) or four years (sandwich): a sandwich course is one where you would spend a period of up to several months with an employer, as part of the course. There are also four- or five-year 'enhanced' MEng degrees. Degree course titles vary – *civil engineering, structural engineering, building engineering, offshore engineering* and *architectural engineering* are some examples. Other courses offer combinations, such as civil and environmental engineering, civil and transportation engineering or civil engineering combined with management, surveying, French or German. If you are aiming to qualify as a chartered engineer, aim for an honours degree, and *check* that any course you are interested in is approved by the Joint Board of Moderators of the Institutions of Civil and Structural Engineers. A list of approved degree courses is available from the Civil Engineering Careers Service.

Courses will usually cover the basic subjects of mathematics, structural engineering, water engineering and geotechnical engineering, as well as surveying, construction materials, use of materials and design projects. Specialist options are often available.

Sponsorships are available from some companies, and the Institution of Civil Engineers offers scholarships to students. Find out about opportunities in *Sponsorship for Students*, published annually by CRAC, or contact the Civil Engineering Careers Service.

Chartered status
Graduates with approved degrees are given structured training by employers, and can take further professional exams and/or a professional review to become chartered engineer of one of the professional bodies, though not all graduate engineers do this. Chartered status proves that an engineer is fully competent and is a recognised qualification worldwide.

Civil engineering surveyor

Civil engineering surveyors support the work of civil engineers throughout the construction industry. Their skills are required both for multi-billion pound projects, such as the Channel Tunnel, and for small-scale operations, such as the renewal of a few hundred metres of service pipeline. **Quantity surveyors** (see earlier section) provide the financial and contractual control; **land surveyors** draw up the plans for a project, showing topographical details and any additional information needed for the construction work, monitoring the progress on a day-to-day basis. There are degree courses in civil engineering surveying or, alternatively, you can qualify through BTEC National and Higher National Diploma courses and through the Institution of Civil Engineering Surveyors' own examination structure.

For a degree course, you need a minimum of two A level passes, or a good grade in a qualification of equivalent standard, such as the Advanced GNVQ in Construction and the Built Environment.

just THE JOB

ROAD MAINTENANCE

> Most people travel by road, either as a driver or a passenger in a car or bus, or as a pedestrian or cyclist. This section is about the road maintenance staff who look after the roads, doing quite a range of jobs: repairing roads and pavements, laying kerbs and drains, maintaining road signs and lighting. People who plan the work or design the roads are technicians or engineers in civil and structural engineering (see previous section).

If you work as a road maintenance operative, you may be employed by the local council. Or you might work for a construction firm which does work for the council on contract.

What the work involves

The team carrying out road repairs needs to be able to:

- follow instructions about the repair to be done;
- direct traffic, put up road signs, traffic lights or safety lamps to make sure that traffic and pedestrians are not put at risk;
- remove the old broken road or paving surface, with a pick and shovel, or a pneumatic drill;
- prepare and mix concrete and other materials, such as tar or asphalt;
- spread concrete or other fillers to level an uneven area;
- cover the road or pavement with the top layer material, such as tar or asphalt. This needs to be spread evenly and rolled;
- clear up when the job is finished.

Depending on the job being done, the team may use equipment like generators, mixers, tractors, excavators and rollers.

There are different grades of workers in road maintenance, such as labourer, roadman or woman, skilled roadman or woman, driver and plant operator. Labourers mainly do the fetching, carrying and unskilled work. Skilled roadmen or women and supervisors have to know more about the technical side of the work, and the use of powered equipment.

What it takes
- You need to be fit and active, and fairly strong – the work can be heavy.

- As you work in a team, it helps a lot if you can get on with other people and work well with them.
- Road maintenance is obviously an outdoor job, and you have to be able to put up with the wet and the cold in winter.
- There is also a lot of dust and dirt, so this would not be suitable work for people with breathing problems.
- Equipment like pneumatic drills makes a tremendous noise. Ear protection is worn, but there is still a risk to hearing.
- It is important to be safety-conscious; aware of your own safety, and also the safety of your workmates and the general public.

GETTING STARTED

These jobs don't usually need any academic qualifications, though GCSEs in basic subjects like maths, English and science can certainly help you to make progress. Young people may be able to start through training with an employer, gaining both practical experience and the chance to study for further qualifications. However, as the work is heavy, there may be a preference for more mature entrants: people who have experience in something similar, like the building industry, will be at an advantage.

Basic training is usually given at the road maintenance depot, and you learn a lot on-the-job from more experienced people. If you prove suitable, you may be able to work towards qualifications such as City & Guilds Roadwork Craft and Advanced Craft certificates, or NVQs up to level 2 in Highways Maintenance. This would help if you were hoping to be promoted to foreman or woman, or supervisor.

JUST THE JOB

WASTES MANAGEMENT

> We live in a world which produces remarkable quantities of waste – domestic rubbish, industrial effluents, radioactive by-products of nuclear power stations and so on. We only notice the work of wastes managers when something goes wrong, such as thousands of dead fish floating in a river or terrible smells from a factory. Controlling and disposing of these wastes is a very technical and complex process, and a high standard of education is required, often to degree level.

Over 100 million tonnes of waste are dealt with each year in this country. 20 million tonnes of it comes from our own homes, over half of which is kitchen waste and paper.

Waste collection
Household waste, and some commercial waste, is collected on a weekly basis. Local councils may run the service, but outside contractors are increasingly being used. Many large industrial companies dispose of their own waste.

Waste disposal
Most household waste is disposed of into landfill sites, where burying the waste allows it to decompose naturally. Several things have to be considered, such as selecting the site, building access roads, landscaping, the effect on the people living nearby, and protecting water sources. Certain industrial and chemical

wastes have to be dealt with in different ways. They can be burnt at very high temperatures, or treated by special processes.

Waste recycling

Nowadays, there is much more interest in recycling wastes, and there are recycling sites all over the country where the public can bring paper, glass, aluminium cans, etc. Local authorities are being encouraged to set up recycling plants where, for example, waste can be burned to produce heat for domestic heating, and electricity for local use.

Wastes management, therefore, may involve knowing something about chemistry, civil engineering, geology and economics. There are other problems to be solved, too, such as the storage and transport of wastes. There are legal and economic considerations, with strict laws on pollution and the ways in which certain wastes can be disposed of.

So there's a lot to waste disposal! This does not mean that a wastes manager needs degrees in engineering, law, chemistry and geology. However, he or she would need to know where to go for specialist advice, and must be able to interpret the results of a study or a consultant's report.

EDUCATION AND TRAINING

The background of wastes managers is very varied. The educational standard required ranges from about five GCSEs at grade C to a degree. If you enter the industry prior to degree level, there are a variety of part-time and distance-learning courses you could take, leading to relevant qualifications. One example is a Higher National Certificate in Wastes Management, designed in conjunction with the Institute of Wastes Management, which is offered at some colleges, for which the minimum entry requirement is generally one A level or the equivalent. The NEBSM Certificate in Supervisory Management, with a Wastes Management option, is another relevant qualification, and the Open University offers a Diploma in Pollution Control.

Graduate entry
If you wish to continue with full-time education to degree level before entering employment, there are a number of degree courses which include modules on waste management or environmental pollution control. Look out for degree courses with titles like *environmental risk management, wastes management and the environment*, or *environment and resource management*, for example. Graduates in subjects varying from civil engineering, environmental science, geology, hydrogeology, biology and chemistry to business management subjects can also enter waste management. The minimum entry requirement for degree courses is two A levels or the equivalent, and supporting GCSEs: check with individual institutions for specific requirements.

Postgraduate courses in waste management and environmental protection are available at a number of universities.

Professional qualifications
Professional training for the industry is overseen by the Institute of Wastes Management. Relevant courses, such as those mentioned above, can count towards full membership of the Institute. The Institute produces lists of relevant courses, including HNC, degree and postgraduate courses, and can advise on routes into the industry.

Managers of sites which require a waste management licence are required by law to hold Certificates of Technical Competence. The route to achieving these is to gain the relevant NVQs which have been developed by the Waste Management Industry Training and Advisory Board (WAMITAB). The NVQs can be achieved while in employment. Previous study, qualifications and experience does not exempt people from the requirement, but can be assessed and accredited towards the NVQs.

just THE JOB

THE WATER INDUSTRY

> When you drink water from your kitchen tap, or pull out the sink plug to let dirty water flow away, you probably don't think of where it's come from or where it goes. Yet over 50,000 people work in the water industry, trying to ensure that the supply is always there and safe to use. Research and engineering jobs are usually done by graduates, but there are also opportunities at craft and technician level for those with fewer qualifications.

The water cycle

The water in your tap has been collected in reservoirs, or may have come from a borehole deep underground. It is filtered, plumbed into mains and piped to your home. The dirty water flows away down sewers to the treatment works, where it emerges cleaned, purified and fit to start the cycle all over again.

The main organisations involved in water resources and supply are the water companies and the Environment Agency. Water supply and sewage disposal are managed by the water companies. River management, flood control, land drainage and pollution monitoring are the responsibility of the Environment Agency. Both the water companies and the Environment Agency are involved in fisheries and leisure and amenity management.

There are ten major water service companies in England and Wales. There are also 22 small companies supplying water only.

The Environment Agency has its headquarters in Bristol but also has eight regional centres. A very high proportion of its staff are involved in technical, scientific and engineering work.

There is a wide range of jobs in the water industry, from water bailiffs to geologists, lorry drivers, laboratory technicians, administrators and accountants. Some jobs are highly specialised, while others are really much the same as in many other commercial and public service organisations.

Jobs in the water industry

The various levels at which staff are employed are described below. Recruitment has been very low over the recent past, and it is unlikely to improve in the near future. You would need to contact employers to find out what the exact situation is in your area.

Operators and craft level jobs

These involve laying and installing water mains and services, inspection of water fittings, operating pumping stations, operating treatment works, concrete construction, transport, clerical work and waterkeeping.

Technicians and supervisory jobs

These involve water supply and treatment, water resources, distribution, sewage treatment and disposal, laboratory work, civil engineering and design, mechanical engineering, finance and general administration.

Degree level

Posts are available for graduate trainees from a wide range of subject areas, including science, engineering and environmentally related subjects, as well as graduates from business, finance and information technology. Graduates are involved in management, research, engineering, financial control and personnel.

TRAINING

All jobs in the water industry involve some training, and the companies have well-organised training for new entrants. This can involve both part-time college courses and in-company training. The Certificate and Assessment Board for the Water Industry (CABWI) offers NVQs at levels 2 and 3 for water services operations and distribution, and NVQs from levels 1 to 3 in laboratory operations. Qualification requirements, entry levels and training offered vary from employer to employer, but a general outline is given below:

Operators in water supply, sewage treatment and land drainage, reservoir attendant. Qualifications are not essential, although they are always helpful to offer. Training would be mainly on-the-job.

Craft and technician trainees in mechanical, electrical and electronic engineering. Qualifications required may be four GCSEs at grade C or an Intermediate GNVQ for some positions, or appropriate A levels, BTEC National or Advanced GNVQ for higher-level technician posts. Training is generally on-the-job with day release, directed towards NVQ levels 2 or 3, or BTEC qualifications.

Graduates in both the scientific and engineering field and in business services would follow a training programme lasting up to three years; the training would be based in the workplace with short specialist courses and possibly the opportunity to study for relevant professional qualifications or postgraduate qualifications.

Adults: note that maturity and previous experience may mean that stated entry requirements can be relaxed.

NAVAL ARCHITECTURE

Naval architects are graduate engineers who are concerned with the design, construction, repair, maintenance and surveying of ships, boats and offshore structures. They work for commercial companies and the Defence Engineering and Science Group. There are also opportunities to work in support of naval architects as technicians, with lower qualifications.

Shipbuilding and repair

Naval architects work in design and drawing offices and in shipyards supervising the progress of a contract. They often move to take up senior technical and management positions in the industry.

Shipping companies

Shipping companies need naval architects to advise on the design of new ships and the repair and maintenance of fleets. There are also specialised technical problems to be solved in such areas as cargo handling.

Consultants

Firms of marine consultants need the skills of naval architects for design and supervision work on contracts. Naval architects who work as consultants have usually had considerable experience.

Ship surveyors

Lloyd's Register of Shipping and the Department of Trade both employ qualified ship surveyors. Lloyds are principally concerned with the condition and structural strength of ships, while the DoT is more concerned with safety regulations.

Defence Engineering and Science Group
This government organisation employs mechanical, electrical and electronic engineers as well as naval architects. They are involved with the design, construction, repair and commissioning of naval vessels, as well as basic research.

Research
Research organisations include the Defence Engineering and Science Group, university departments and Lloyd's Register of Shipping.

Equipment manufacturers
Naval architects are employed on such products as propulsion systems and controls.

Offshore rig fabrication and siting
Technological advances associated with deep-sea operations have increased the demand for naval architects in the oil and gas industries.

Small craft and yacht builders
Naval architects contribute to the design and building of smaller vessels as well as large ships.

EDUCATION AND TRAINING

A fully qualified naval architect is a **chartered engineer**. He or she will have taken an honours degree in an approved engineering subject, followed by two years of training and a further two years doing a responsible job in a chosen field of specialisation. There are degree courses in naval architecture. Graduates in other engineering disciplines who have appropriate experience are also acceptable to the Royal Institution of Naval Architects.

The minimum entry qualifications for a degree course in naval architecture or engineering are usually three good A levels (to include mathematics and physics) plus supporting GCSEs. BTEC qualifications at National Diploma and Higher National level are also acceptable. Applicants offering an Advanced GNVQ in science *plus* an additional A level will be considered on their individual merits by admissions tutors.

Most student naval architects go straight into higher education from school, but it is also possible to start as a **technician** and, with good BTEC qualifications, move on to a degree course. Entry requirements for trainee technicians vary, but are usually three or four GCSEs at grade C including maths, science (with a high physics content) and English.

Adults: suitably qualified adults may train in naval architecture, but it is unlikely that exemptions would be granted from the qualifications specified, because of the scientific and technical knowledge needed.

just THE JOB

SHIPBUILDING & BOATBUILDING

Shipbuilding is the business of building large vessels, mainly of steel. *Boatbuilding* is the term used for the construction of small vessels from wood, steel or fibre-reinforced plastic. There are career opportunities in both building and repair work for workers at every level, from craftsmen and women to graduate designers and engineers.

Shipbuilding

The British shipbuilding industry developed in areas where deep water and sheltered anchorages could be easily supplied with steel and the fuel for marine engines, which was originally coal. The most suitable localities are still in the north of England and Scotland, on rivers such as the Tyne, the Clyde and the Humber and the peninsula of Barrow in Cumbria. There are many successful smaller boatbuilding yards in the south of England.

Ship repairing, usually a smaller-scale business, is carried out in some additional areas such as Falmouth in Cornwall and other ports round the coast. The industry has suffered serious decline in the last twenty years, but appears to be stabilising, with approximately fifty significant companies remaining in business. Skills learned for shipbuilding and repair are, in some cases, readily transferable to other engineering sectors and, of course, the reverse is also true.

Modern shipbuilding is a highly complex process using

computer-aided design and manufacture. There are various levels of employment, depending on qualifications and experience.

Craft trainee
There are many specialised craft jobs involved in turning the design team's work into a ship. Skills include welding, steel cutting and bending, marking out shapes, assembling sections, pipework, installing engines and doing internal fittings. Training is usually on-the-job, with day release or block release for NVQ modules in individual skills. Qualifications for entry are usually good grades in maths and science GCSEs.

Engineering technician
Technicians are employed on such work as drafting, estimating, testing, quality control, surveying, computing and production control.

Craftsmen and women may move up to technician level by gaining further qualifications, such as BTEC National Certificate/NVQ level 3. Entry at technician level is also possible with three or four good GCSEs, preferably including maths and science, or an Intermediate GNVQ.

Many shipbuilders and ship repair yards now offer Modern Apprenticeships. These are designed for 16 to 19 year olds, with at least four GCSEs at grade C. The Apprenticeships last for two or three years and lead to NVQs at level 3, at least.

Incorporated and chartered engineers
Engineers are highly qualified people concerned with the design, building, planning, surveying and repair of ships. Degree-level courses, which can lead to chartered engineer status, and Higher National Diploma and Certificate courses leading to incorporated engineer status, are available at a number of institutions. A list of these universities and colleges can be

obtained from the Royal Institution of Naval Architects (see Further Information section).

EMPLOYERS

The small amount of continuing work in shipbuilding is largely confined to the traditional geographical areas mentioned earlier. Marine engines and components are made by various large concerns often involved in other engineering work.

Boatbuilding

Boatbuilding means making pleasure and small commercial craft up to about 50 metres in length. There are also makers of auxiliary equipment such as sails, engines and boat fittings. Boatbuilders work in a variety of materials: wood, iron, steel, aluminium, copper, fibre-reinforced plastics (FRP), cloths and paints. There are about 1200 boatbuilders, repairers and small companies involved in other aspects of the industry. Over half of British-made boats are currently exported.

There is still a high degree of craft skill in the industry. As a craftsman or woman you could be a boatbuilder, boat-fitter, electrician, joiner, painter, plumber, rigger or sailmaker. In the technical field you could train to be a draughtsman or woman, boat designer or boatyard manager. There are also jobs in administration, management, accounts, and in clerical and secretarial work.

TRAINING

Craft training can be either on- or off-the-job. The trade federation, the British Marine Industries Federation (which represents the firms building and repairing small ships and pleasure boats), has developed NVQs at levels 2 and 3, and a Modern Apprenticeship. It would, of course, be possible to learn a trade such as plumbing, painting or joinery and subsequently go into boatbuilding. Details about entering training through employment can be obtained from your careers service. There are a number of universities and colleges which offer relevant courses.

FOR FURTHER INFORMATION

GENERAL

Building Employers' Confederation – 82 New Cavendish Street, London W1M 8AD. Tel: 0171 580 5588.

Chartered Institute of Building – Englemere, Kings Ride, Ascot, Berks SL5 8BJ. Tel: 01344 23355.

Construction Industry Training Board – Careers Advisory Service, Bircham Newton, King's Lynn, Norfolk PE31 6RH. Tel: 01553 776677.

JTL – Head Office: South Block, Central Court, Knoll Rise, Orpington, Kent BR6 0JA. Tel: 01689 891676.

Working in Construction, published by COIC.

CARPENTER & JOINER

Guild of Master Craftsmen – 166 High Street, Lewes, East Sussex BN7 1XU. Tel: 01273 478449.

Institute of Carpenters – Central Office, 35 Hayworth Road, Sandiacre, Nottingham NG10 5LL. Tel: 0115 9490641. The Institute publishes a brochure entitled *Take up a Natural Career in Wood*.

ROOFING

Institute of Roofing – 24 Weymouth Street, London W1N 3FA. Tel: 0171 436 0103.

Mastic Asphalt Council – Lesley House, 6 & 8 Broadway, Bexleyheath, Kent DA6 7LE. Tel: 0181 298 0411.

National Federation of Roofing Contractors – 24 Weymouth Street, London W1N 4LX. Tel: 0171 436 0387.

PLASTERING
Federation of Plastering and Drywall Contractors – 18 Mansfield Street, London W1M 9FG. Tel: 0171 580 5404.

PAINTING AND DECORATING
British Decorators' Association – 32 Coton Road, Nuneaton, Warwickshire CV11 5TW. Tel: 01203 353776.

National Federation of Painting and Decorating Contractors – 18 Mansfield Street, London W1M 9FG. Tel: 0171 580 5404.

TILING
National Master Tile Fixers Association – 39 Upper Elmers End Road, Beckenham, Kent BR3 3QY. Tel: 0181 663 0946.

PLUMBING
Joint Industry Board for Plumbing – Brook House, Brook Street, St Neots, Huntingdon, Cambridgeshire PE19 2HW. Tel: 01480 476925.

National Association of Plumbing, Heating and Mechanical Services Contractors – 14–15 Ensign Business Centre, Westwood Business Park, Westwood Way, Coventry CV4 8JA. Tel: 01203 470626.

ELECTRICAL INSTALLATION AND MAINTENANCE
Engineering Careers Information Service – EnTra, 41 Clarendon Road, Watford, Herts WD1 1HS. Freephone: 0800 282167.

JTL – Head Office, South Block, Central Court, Knoll Rise, Orpington, Kent BR6 0JA. Tel: 01689 891676.

HEATING, VENTILATION & AIR CONDITIONING
Building Engineering Services Training Limited (BEST) – 3 Mill Court, 51 Mill Street, Slough, Berkshire SL2 5DA. Tel: 01753 531188.

Chartered Institute of Building Services Engineers – 222 Balham High Road, London SW12 9BS. Tel: 0181 675 5211.

HVCA (Heating and Ventilating Contractors Association) Education and Training Department – ESCA House, 34 Palace Court, London W2 4JG. Tel: 0171 229 2488. Can provide leaflets describing entry requirements, training and prospects.

REFRIGERATION
Engineering Services Training Trust Ltd (ESTTL) – ESCA House, 34 Palace Court, London W2 4JG. Tel: 0171 229 2488.
Institute of Refrigeration – Kelvin House, 76 Mill Lane, Carshalton, Surrey SM5 2JR. Tel: 0181 647 7033.

STEEPLEJACK
You will find local steeplejack firms listed in *Yellow Pages*.

Construction Industry Training Board (Steeplejacks) – Belton Road Industrial Estate, Prince William Road, Loughborough, Leicester LE11 0GU. Tel: 01509 610266 (ask for Mike Feneley).

STONE CLEANING
Stone Federation Great Britain – 18 Mansfield Street, London W1M 9FG. Tel: 0171 580 5404. Can provide occupational information for people interested in working as a stone cleaner.

You will find local firms listed in *Yellow Pages* under 'Stone cleaning and restoration'.

DEMOLITION
National Demolition Training Group – CITB, Newton Training Centre, Bircham Newton, King's Lynn, Norfolk PE31 6RH. Tel: 01553 776677.

BUILDING TECHNICIANS, TECHNOLOGISTS & MANAGERS
Institute of Clerks of Works of Great Britain – 41 The Mall, London W5 3TJ. Tel: 0181 579 2917.
Royal Institution of Chartered Surveyors – Surveyor Court, Westwood Way, Coventry CV4 8JE. Tel: 0171 222 7000 or 01203 694757.

SURVEYING

Architects and Surveyors Institute – St Mary House, 15 St Mary Street, Chippenham, Wiltshire SN15 3WD. Tel: 01249 444505.

Association of Building Engineers – Jubilee House, Billing Brook Road, Weston Favell, Northampton NN3 8NW. Tel: 01604 404121.

Chartered Institute of Building – Englemere, Kings Ride, Ascot, Berks SL5 8BJ. Tel: 01344 23355.

Institution of Civil Engineering Surveyors – Newholme, 26 Market Street, Altrincham, Cheshire WA14 1PF. Tel: 0161 928 8074.

Royal Institution of Chartered Surveyors – Surveyor Court, Westwood Way, Coventry CV4 8JE. Tel: 0171 222 7000 or 01203 694757.

Society of Surveying Technicians – Surveyor Court, Westwood Way, Coventry CV4 8JE. Tel: 0171 222 7000 or 01203 694757.

Building a Career has a section on building surveying work, and is available from the Construction Careers Service, CITB, Bircham Newton, King's Lynn, Norfolk PE31 6RH. Tel: 01553 776677.

ARCHITECTURE

Architects Registration Council of the UK (ARCUK) – 73 Hallam Street, London W1N 6EE. Tel: 0171 580 5861. ARCUK can supply information on application procedures, and a leaflet *United Kingdom Architects in Europe*.

Architects and Surveyors Institute – St Mary House, 15 St Mary Street, Chippenham, Wiltshire SN15 3WD. Tel: 01249 444505. Can provide careers information leaflets.

British Institute of Architectural Technologists – 397 City Road, London EC1V 1NE. Tel: 0171 278 2206. Produces the useful leaflets *Education and Training for a Career in Architectural Technology* and *Qualify for a Professional Career as an Architectural Technologist*, and a directory of courses approved by BIAT.

Royal Institute of British Architects – 66 Portland Place, London W1N 4AD. Tel: 0171 580 5533.

Schools of Architecture, available from RIBA Publications Ltd, Finsbury Mission, Moreland Street, London EC1V 8VB. Tel: 0171 251 0791.

A Career in Architecture and *Advice to Employers of Students* are free booklets from the Education and Professional Development Department at the RIBA (address as above).

Careers in Architecture, published by Kogan Page.

Degree Course Guide: Architecture, Planning and Landscape Architecture, published by CRAC/Hobsons Press.

Architecture, Landscape Architecture and Town and Regional Planning, an AGCAS booklet (graduate careers information) available in careers libraries or from CSU, Armstrong House, Oxford Road, Manchester M1 7ED. Tel: 0161 236 9816, ext 250/251.

CIVIL & STRUCTURAL ENGINEERING

Association of Consulting Engineers – Alliance House, 12 Caxton Street, London SW1H 0QL. Tel: 0171 222 6557. Internet: http://www.acenet.co.uk.

Civil Engineering Careers Service – 1 Great George Street, London SW1P 3AA. Tel: 0171 222 7722. Internet: http://www.ice.org.uk.

Construction Industry Training Board – Bircham Newton, King's Lynn, Norfolk PE31 6RH. Tel: 01553 776677.

Institution of Civil Engineering Surveyors – Newholme, 26 Market Street, Altrincham, Cheshire WA14 1PF. Tel: 0161 928 8074.

Institution of Civil Engineers – 1 Great George Street, Westminster, London SW1P 3AA. Tel: 0171 222 7722.

Institution of Structural Engineers – 11 Upper Belgrave Street, London SW1X 8BH. Tel: 0171 235 4535.

WASTES MANAGEMENT

Environmental Services Association – 154 Buckingham Palace Road, London SW1W 9TR. Tel: 0171 824 8882.

Institute of Wastes Management – 9 Saxon Court, St Peter's Gardens, Northampton NN1 1SX. Tel: 01604 20426.

Waste Management Industry Training and Advisory Board (WAMITAB) – PO Box 176, Northampton NN1 1SB. Tel: 01604 231950.

THE WATER INDUSTRY

Board for Education and Training in the Water Industry (BETWI) – 1 Queen Anne's Gate, London SW1H 9BT. Tel: 0171 957 4524.

Certificate and Assessment Board for the Water Industry (CABWI) – 1 Queen Anne's Gate, London SW1H 9BT. Tel: 0171 957 4523.

Environment Agency – Rio House, Waterside Drive, Aztec West, Almondsbury, Bristol BS12 4UD. Tel: 01454 624400.

Institution of Water Officers – Heriot House, 12 Summerhill Terrace, Newcastle upon Tyne NE4 6EB. Tel: 0191 230 5150. Deals with enquiries relevant to membership, but does not provide careers information.

Water Companies Association – 1 Queen Anne's Gate, London SW1H 9BT. Tel: 0171 222 0644.

Water Services Association – 1 Queen Anne's Gate, London SW1H 9BT. Tel: 0171 957 4567.

Anglian Water – Ambury Road, Huntingdon PE18 6NZ. Tel: 01480 443000.

Dwr Cymru – Plas-y-Ffynnon, Cambrian Way, Brecon, Powys LD3 7HP. Tel: 01874 623181.

East of Scotland Water Authority – Head Office, Pentland Gait, Calder Road, Edinburgh EH11 4HJ. Tel 0131 445 4141.

North of Scotland Water Authority – Caledonia House, 63 Academy Street, Inverness IV1 1LU. Tel: 01463 245 400.

North West Water – Dawson House, Great Sankey, Warrington, Cheshire WA5 3LW. Tel: 01925 234000.

Northern Ireland Water Services – Northland House, 3 Frederick Street, Belfast BT1 2NS. Tel: 01232 244711.

Northumbrian Water – Abbey Road, Pity Me, Durham DH1 5FJ. Tel: 0191 383 2222.
Severn Trent Water – 2297 Coventry Road, Birmingham B26 3PU. Tel: 0121 722 4000.
South West Water – Peninsula House, Rydon Lane, Exeter EX2 7HR. Tel: 01392 446688.
Southern Water – Southern House, Yeoman Road, Worthing BN13 3NX. Tel: 01903 264444.
Thames Water – Nugent House, Vastern Road, Reading RG1 8DB. Tel: 01734 591159.
Wessex Water – Wessex House, Passage Street, Bristol BS2 0JQ. Tel: 0117 929 0611.
West of Scotland Water Authority – 419 Balmoral Road, Glasgow G22 6NU. Tel: 0141 355 5333.
Yorkshire Water – West Riding House, 67 Albion Street, Leeds LS1 5AA. Tel: 0113 244 8201.

NAVAL ARCHITECTURE

Defence Engineering and Science Group – Graduate Recruitment Office, Room 136, Pinesgate East, Lower Bristol Road, Bath BA1 5AB. Tel: 01225 449106.
Institute of Marine Engineers – The Memorial Building, 76 Mark Lane, London EC3R 7JN. Tel: 0171 481 8493.
Royal Institution of Naval Architects – 10 Upper Belgrave Street, London SW1X 8BQ. Tel: 0171 235 4622. The Institution produces a useful booklet, *Careers in Naval Architecture,* which describes all aspects of the work of a naval architect and the entry routes to relevant university and college courses.

SHIPBUILDING & BOATBUILDING

British Marine Industries Federation – Meadlake Place, Thorpe Lea Road, Egham, Surrey TW20 8HE. Tel: 01784 473377.
Institute of Marine Engineers – The Memorial Building, 76 Mark Lane, London EC3R 7JN. Tel: 0171 481 8493.

Marine and Engineering Training Association – Rycote Place, 30–38 Cambridge Street, Aylesbury, Bucks HP20 1RS. Tel: 01296 434943.

Royal Institution of Naval Architects – 10 Upper Belgrave Street, London SW1X 8BQ. Tel: 0171 235 4622.

Shipbuilders and Shiprepairers Association – 33 Catherine Place, London SW1E 6DY. Tel: 0171 828 0933.